KRV JUN 25 '91

SIMPLE
ASTRONOMY

SIMPLE ASTRONOMY

by Iain Nicolson
illustrated by Don Pottinger

Charles Scribner's Sons · New York

Printed in the United States of America
Library of Congress Catalog Card Number 73–11573
ISBN 0-684-13640-6

CONTENTS

CHAPTER ONE

THE MOVING SKY

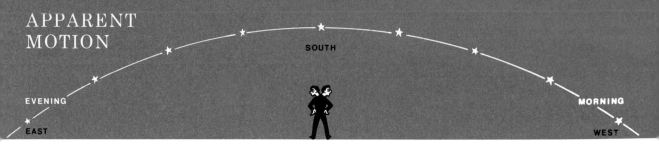

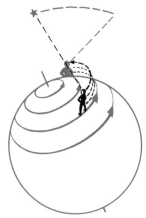

Motion of an observer on the earth

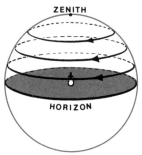

Apparent motion of stars at the poles

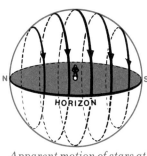

Apparent motion of stars at the equator

From the earliest times, the sky-watcher must have been aware of the regular cyclic motions of the heavens. For, in the northern hemisphere, sun, moon and stars alike rise in the east and move westwards across the sky, rising higher above the horizon until they are due south ("Culmination"). Thereafter, they sink back towards the horizon, finally setting in the west. Day after day and night after night the same cycle is repeated. There are other motions, too. The moon changes its position relative to the stars, making a circuit of the sky once in about four weeks, and the particular stars visible at night change as the seasons progress, but always the same patterns return after the space of a year. In ancient times it seemed perfectly reasonable to conclude that the sky was a dome set above the earth, and that the earth lay at the centre of the universe. Later, the early Greek astronomers came to the conclusion that the stars were fixed to a sphere which rotated around the earth.

Today we know that the earth is a globe rotating on its axis from west to east once every twenty-four hours, and that the stars lie at great distances from us. They do not rotate around the earth, but because an observer on the earth's surface is being carried round from west to east by the earth's rotation, it is easy to see that the stars will appear to move across the sky from east to west. Nevertheless, it's very convenient in some ways to retain the notion that the stars are fixed to a giant sphere called the *celestial sphere*. If we imagine the earth's axis extending north and south from the poles, outwards through space, then the points at which it reaches the celestial sphere are the north and south *celestial poles*. The sphere is imagined to rotate around this axis, and hence around the earth, once a day. The *celestial equator* is also analogous to that of the earth.

If we were at the north pole of the earth we would find that the north celestial pole was vertically overhead and that the stars would not rise and set, but would move around the sky in circles parallel to the horizon. Furthermore, half of the stars on the celestial sphere would be visible all the time, and the other half never be seen. If, however, we went to the equator, then the celestial poles would be on the horizon, and all the stars could be seen at one time or another.

8

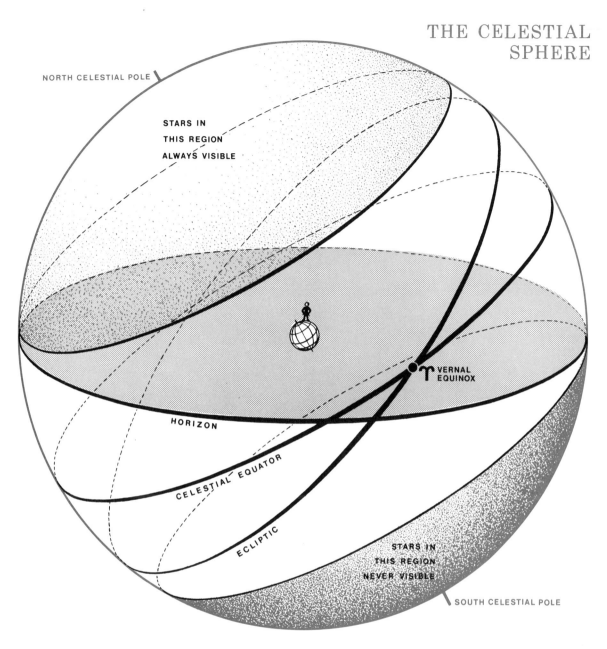

NORTH CELESTIAL POLE

STARS IN
THIS REGION
ALWAYS VISIBLE

VERNAL
EQUINOX

HORIZON

CELESTIAL EQUATOR

ECLIPTIC

STARS IN
THIS REGION
NEVER VISIBLE

SOUTH CELESTIAL POLE

Most of us live neither at the equator nor at a pole, so we find that most stars rise and set. Some stars are permanently hidden below our horizon, and certain others which lie close to the pole of the sky are permanently visible. Such stars are known as *circumpolar* stars. We regard the stars for many purposes as being more or less fixed on the celestial sphere. Consequently, just as we can map the surface of the earth, so we can draw maps of the positions of the stars on the celestial sphere. The only difference in the case of the stars is that we are inside the sphere looking out.

Star chart showing half the visible sky as seen at a particular time and place

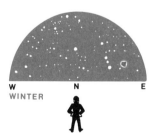

W   N   E
WINTER

9

SEASONAL CHANGES

SUMMER CHART

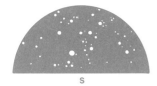

WINTER CHART

In addition to the daily rotation of the celestial sphere there are seasonal changes in the positions of stars in the sky. The reason for this isn't hard to see. As we know, the earth moves round the sun in a nearly circular path once a year and as we can see above, an observer on the daylight side of the earth will see the sun projected against the distant background stars. As the earth moves, so the sun appears to move relative to the background stars, tracing out a circle on the celestial sphere known as the *ecliptic*. By the same token an observer on the night side of the earth will see different parts of the celestial sphere as the earth moves around the sun. Because of this effect, star charts must be drawn for a particular season as well as a particular time. The stars visible in summer, apart from the circumpolar ones, will be quite different from those visible in winter.

10

One of the earliest practical applications of astronomy was in the field of navigation. The stars visible in the sky depend upon one's position on the surface of the earth, and so observations of stars may be used to determine position. Position on the earth is measured in terms of latitude (measured north or south of the equator) and longitude (measured east or west of Greenwich), while star positions are measured relative to the celestial equator and poles of the earth. Astro-navigation depends upon an accurate knowledge of the positions of stars, and of time, and it was primarily to make precise measurements of these quantities that the Royal Greenwich Observatory was founded in 1675.

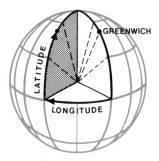

Position measurement on the earth

Position in the sky can be measured in several ways, two of which are shown on the right. The observer may denote the star's position by *altitude*, its angular height above the horizon, and *azimuth*, its angular distance from the north point of the horizon. As the night progresses, the movement of the sky causes both altitude and azimuth to change. The point on the sky directly overhead is the *zenith*, and the circle passing through the north point, zenith and south point is the observer's *meridian*. It is used in the second system where position is given by *declination*, the angular distance of the star from the celestial equator, and *hour angle*, the angle between the meridian, the north celestial pole and the star. Hour angle is measured clockwise from the meridian, and increases uniformly as time passes, while declination remains constant.

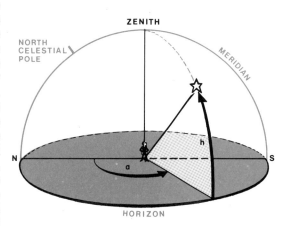

Altitude (h) and azimuth (a) for a star

As its name suggests, the hour angle is closely related to time, and is measured not in degrees but in hours, minutes and seconds. If we say that 24 hours is the time taken for the sun to cross the meridian, move round the sky and return to the meridian, then 24 hours is equivalent to 360°. When the sun is due south on the meridian at noon its hour angle is zero, six hours later it has moved through 90° and its hour angle is 6 hours, and so on. Thus, local solar time is simply the hour angle of the sun plus 12 hours. *Greenwich Mean Time* is solar time measured at the longitude of Greenwich, and corrected for variations in the motion of the sun. If we measure time by the hour angles of stars we obtain *sidereal time*.

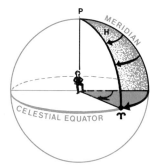

Sidereal time is the hour angle of a point (γ), the vernal equinox, which is fixed relative to the stars on the celestial sphere

11

PLANETARY MOTIONS

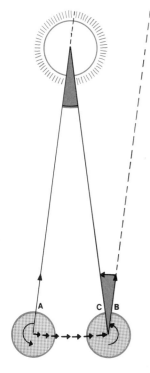

A consequence of the motion of the earth around the sun is that the sidereal day and solar day are not quite the same length, as illustrated opposite. In one sidereal day the observer on the earth rotates through 360° relative to the stars, i.e., he moves from A to B. If the sun was on the meridian at A, then, because the earth has moved relative to the sun in a day, the sun will not quite be on the meridian for the observer at B. The earth will have to rotate a little further to bring the observer to C before the sun reaches his meridian. This extra rotation takes about four minutes, so the sidereal day is some four minutes shorter than the solar day, and although there are $365\frac{1}{4}$ solar days in the year the earth actually rotates $366\frac{1}{4}$ times.

Planets betray their presence in the sky by their motions, which depend on whether or not they are closer to the sun than the earth. Planets closer to the sun are known as *inferior planets*, and are always seen fairly close to the sun, either in the evening sky after sunset or the morning sky before sunrise. When such a planet is in line with the sun and at its closest to us, it is at *inferior conjunction*. *Superior conjunction* occurs when the planet is at its greatest distance. Its maximum angular distance from the sun is *greatest elongation*. Planets further from the sun than the earth are *superior planets*. When at their closest to us they are directly opposite the sun in the sky and so are visible due south at midnight. They are then said to be at *opposition*. Superior planets can also reach superior conjunction, but can clearly never be seen at inferior conjunction.

Movement of the earth relative to the sun in one day

Inner planets

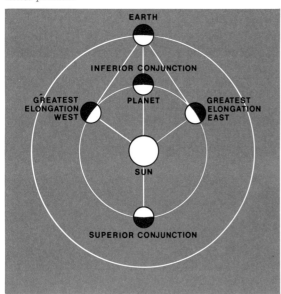

Outer planets

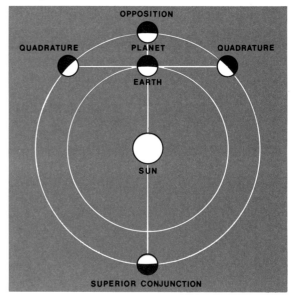

CHAPTER TWO

MEASURING THE UNIVERSE

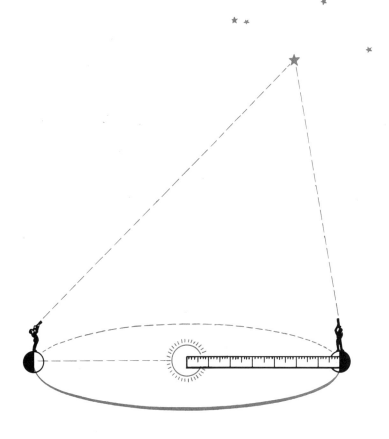

DISTANCE AND SIZE

EARTH   EDGE OF
          THE SUN

Relative sizes of the earth and a segment of the sun

The earth is by no means the most important body in the universe. We now know that it is merely a planet, a member of a system of planets, known as the *solar system*, each of which revolves around the sun. Compared with the sun the earth is very small, only one hundredth of its diameter, and one millionth of its volume. However, the sun itself is a star, of no particular significance among the thousands of millions of stars which make up our star system or *galaxy*. Even our galaxy is not unique, for present-day telescopes can detect thousands of millions of similar systems, some of them so distant that the light we are now receiving from them set out from these galaxies thousands of millions of years ago.

The scale of the universe is so enormous that it's quite impossible to visualise the distances and sizes involved; the human imagination simply cannot cope. We can, however, represent the sun by a football, in which case the earth would be a ball-bearing, 2·5 millimetres in diameter and located some 25 metres distant, while the moon would lie about 7 centimetres from the earth. The giant planet Jupiter could be denoted by a table-tennis ball some 125 metres from the sun, and Pluto, the most distant member of the solar system, would be about 2 kilometres from our "sun". On this scale the nearest star would be 4,100 miles or 6,500 km. distant. Even the model begins to stretch the imagination, yet if we wish to represent our galaxy accurately we should need a model as large as the actual distance from the earth to the sun! What is really important is not an understanding of absolute values of size and distance, but an appreciation of relative sizes and distances. On the facing page our horizons are widened step-by-step by treating the distance scale of the universe in five different stages where each stage shows an area of the universe roughly ten thousand times larger than the preceding one.

Stage one represents the region of space occupied by the earth and the moon. Over this range miles or kilometres are still convenient units to use, and the moon lies some 238,000 miles (384,000 km.) from the earth. Stage two represents the solar system where a common unit of distance is the *astronomical unit* (A.U.). (1 A.U.=93 million miles (150 million km.). The solar system radius is about 40 A.U. Stage three depicts the nearby stars out to a range of 10 *light-years*, where one light year is the distance travelled by a ray of light in one year. Since light moves at 186,000 miles (300,000 km.) per second, 1 light year equals about 6 million million miles (10 million million km.). Stage four represents the scale of our galaxy, some 100,000 light-years across, and stage five the boundaries of the observable universe some 10 thousand million light-years distant.

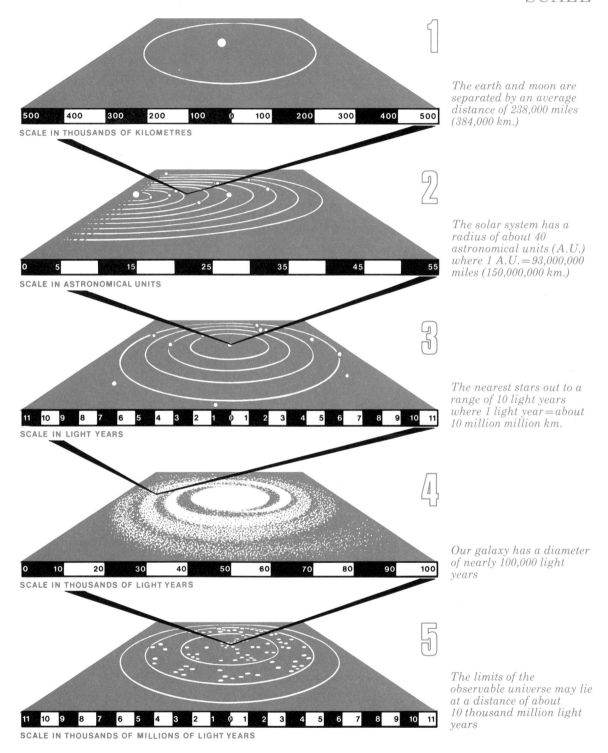

The earth and moon are separated by an average distance of 238,000 miles (384,000 km.)

SCALE IN THOUSANDS OF KILOMETRES

The solar system has a radius of about 40 astronomical units (A.U.) where 1 A.U.=93,000,000 miles (150,000,000 km.)

SCALE IN ASTRONOMICAL UNITS

The nearest stars out to a range of 10 light years where 1 light year=about 10 million million km.

SCALE IN LIGHT YEARS

Our galaxy has a diameter of nearly 100,000 light years

SCALE IN THOUSANDS OF LIGHT YEARS

The limits of the observable universe may lie at a distance of about 10 thousand million light years

SCALE IN THOUSANDS OF MILLIONS OF LIGHT YEARS

EARTH-MOON
RELATIONSHIP

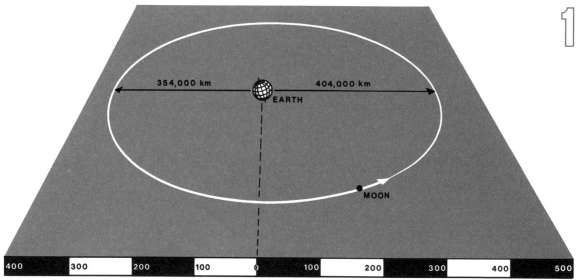

SCALE IN THOUSANDS OF KILOMETRES

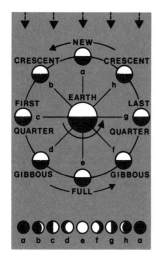

PHASES OF THE MOON
(a) new moon; (b) crescent;
(c) first quarter; (d) gibbous;
(e) full moon; (f) gibbous;
(g) third quarter;
(h) crescent

The moon is by far our nearest celestial neighbour and moves around the earth in a path, or *orbit*, which is very nearly circular, having an average radius of 238,000 miles or 384,000 km. Relative to the stars, it completes one circuit of the earth in 27·3 days, this being its *sidereal period*. The moon shines solely by reflected sunlight and, as it moves round the earth, we see a varying proportion of the illuminated side. In other words, the moon exhibits a cycle of phases. As we can see in the diagram, when the moon is more or less in line with the sun the dark side is turned towards us and we have a "new moon". As the moon moves away from the sun in the sky so the proportion of the illuminated side which we see increases until the moon is opposite the sun in the sky, when the side we see is fully illuminated and we have "full moon". Thereafter the moon approaches the sun again, and the phase decreases to "new moon" once more.

The time taken for the moon to complete its cycle of phases is longer than its sidereal period, since in the time taken for it to travel round the earth has moved some way round the sun. The moon will not return to new moon until 29½ days have elapsed, this being the *synodic period* of the moon. Occasionally, at new moon the moon will pass directly in front of the sun, blocking out its light and giving rise to an eclipse of the sun. Conversely, at full moon an eclipse of the moon can occur if it enters the shadow of the earth. To help retain an impression of scale it is worth remembering that light takes 1¼ seconds to reach the moon.

16

The sun and planets, together with minor planets, comets and other material to be mentioned later make up the solar system. The motions of planets in the sky were explained in 1609 by Johannes Kepler who deduced that each planet, including the earth, moves round the sun in an elliptical path with the sun located at a focus of the ellipse. Furthermore he showed that they move in such a way that the line joining a planet to the sun sweeps out equal areas of space in equal times; in other words, planets move faster when closer to the sun than when further away. For example, the earth is closer to the sun during the southern hemisphere summer than it is during the northern hemisphere summer. Since it moves faster when closer to the sun, summer is slightly shorter and warmer in the southern hemisphere than it is in the northern hemisphere.

The nine major planets are, in order from the sun, Mercury, Venus, Earth, Mars, Jupiter, Saturn, Uranus, Neptune and Pluto. All but the last three can be seen without optical aid, and so have been known since ancient times. The four innermost planets are somewhat similar to the earth in size and structure, and are known as the "terrestrial planets". The next four are huge gaseous bodies, known as the "Jovian" or "Giant" planets, whereas Pluto is something of a mystery, but appears to be a small solid body comparable in size with Mercury. Distances from the sun range from 36 million miles (58 million km.) for Mercury to 3,660 million miles (5,900 million km.) for Pluto, and orbital periods from 88 days to 248 years respectively. It takes a light ray about five hours to travel from the sun to Pluto.

KEPLER'S LAWS OF
PLANETARY MOTION

*Each planet moves round
the sun in an ellipse in
such a way that if it moves
from 1 to 2 in unit time
it will also move from 3 to 4
in the same time and
area A will equal area B*

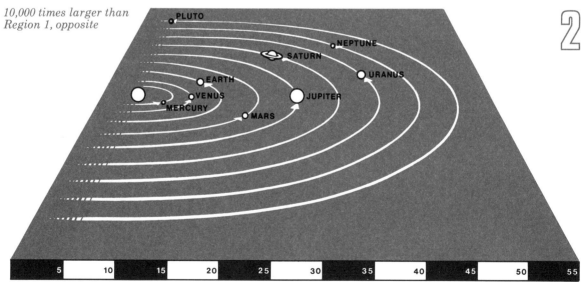

*10,000 times larger than
Region 1, opposite*

SCALE IN ASTRONOMICAL UNITS

(relative distances of planets not to scale)

*10,000 times larger
than Region 2*

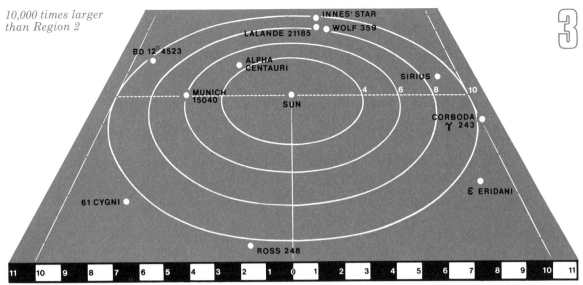

SCALE IN LIGHT YEARS

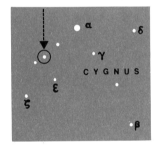

Position of 61 Cygni

The distance to a star was first accurately measured by Friedrich Bessel in 1838 by means of the method of *parallax*, described on page 43. The star concerned was in the constellation of Cygnus, and the distance ascribed to it just over ten light years. This is not the nearest star to our sun but only eight are known to lie closer to us than ten light years, although another six at least lie in the range ten to eleven light years. The closest star is a faint red one known as Proxima Centauri, which although lying at a distance of only 4·2 light years is too faint by far to be seen without optical aid. Its real brightness (not to be confused with the apparent brightness it seems to have because of its distance) is only one ten thousandth that of the sun. It is associated with Alpha Centauri, a bright double star whose distance is 4·3 light years.

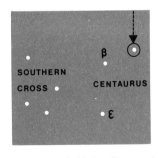

Position of Alpha Centauri

Stars vary enormously in brightness. Of these eight nearest stars, the difference in brightness between the brightest and the faintest is a factor of a million, and only two are bright enough to be seen without a telescope. One of these is Alpha Centauri and the other Sirius, which appears to be the brightest star in the sky and is inherently twenty-six times brighter than the sun. If we extend our range to include the twenty nearest stars, we find only seven of these are visible to the naked eye, and only three are inherently brighter than the sun. This is not a representative sample of stars, but it does look as though there are more stars fainter than the sun than brighter.

18

The sun is a member of a giant star system known as our *galaxy*, which we know contains approximately one hundred thousand million stars of all kinds distributed in a disc-like structure, the diameter of which is nearly one hundred thousand light years. As well as stars our galaxy contains clouds of gas and dust. One of the most difficult tasks which astronomers have undertaken in the last two centuries has been the determination of the structure and dimensions of our star system. Some of the best early work was done by the pioneer observer William Herschel in the late eighteenth and early nineteenth centuries, but recent research has included both optical and radio astronomical results to establish that the sun lies some 30,000 light years from the centre of the galaxy. It's also been ascertained that the sun, solar system and nearby stars alike move round the centre of the galaxy in a span of about 225 million years, a period of time sometimes known as the "Cosmic Year".

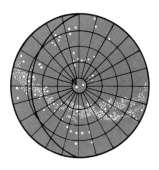

Appearance of the Milky Way

The misty band of light, known as the Milky Way, which can be seen across the sky on a clear, moonless night, is seen telescopically to be due to the combined light of millions of faint stars. It is readily explained in terms of the disc shape of the galaxy. If we look away from the plane of the galaxy disc then we are looking through a thin part of the system and will see relatively few stars, but if we look along the plane we will see vast numbers of stars apparently crowded together and giving rise to the Milky Way.

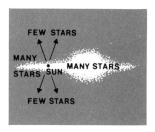

Explanation of the Milky Way

10,000 times larger than Region 3

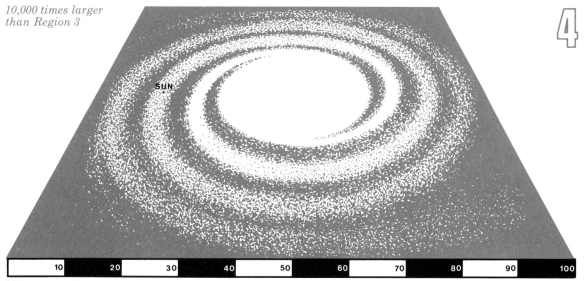

SCALE IN THOUSANDS OF LIGHT YEARS

THE VISIBLE UNIVERSE

100,000 times larger than Region 4

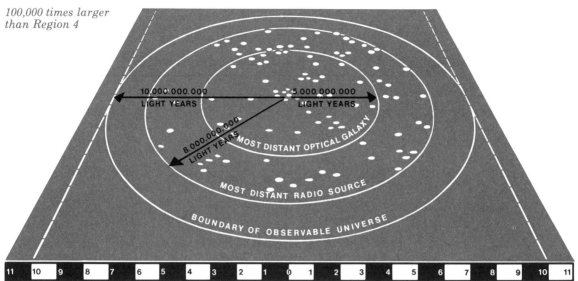

10,000,000,000 LIGHT YEARS

5,000,000,000 LIGHT YEARS

8,000,000,000 LIGHT YEARS

MOST DISTANT OPTICAL GALAXY

MOST DISTANT RADIO SOURCE

BOUNDARY OF OBSERVABLE UNIVERSE

| 11 | 10 | 9 | 8 | 7 | 6 | 5 | 4 | 3 | 2 | 1 | 0 | 1 | 2 | 3 | 4 | 5 | 6 | 7 | 8 | 9 | 10 | 11 |

SCALE IN THOUSANDS OF MILLIONS OF LIGHT YEARS

The Andromeda Galaxy and its location

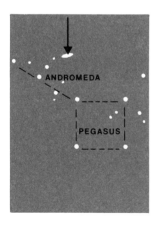

ANDROMEDA

PEGASUS

Within range of existing telescopes there are thousands of millions of galaxies, broadly similar to our own. Our galaxy is a member of a small cluster of galaxies, known as the *Local Group* and numbering about twenty assorted star systems, of which our galaxy is one of the largest. The best-known member of the local group is the Andromeda Galaxy, a star system slightly larger than our own, but very similar and lying at a distance of about 2·2 million light years. Even at this great distance it is sufficiently luminous to be visible to the naked eye on a clear night as a faint misty patch in the sky.

The many millions of galaxies which lie beyond the local group also tend to group into clusters of up to several hundred members. The most remarkable thing about these galaxies is that, almost without exception, they are moving away from us, and the more distant they are the faster they seem to be moving. The most distant galaxies visible in optical telescopes are at ranges of about 5,000 million light years, but radio telescopes seem to be detecting radiation from objects which may be as much as 8,000 million light years distant. If the expansion of the universe of galaxies holds true even at the limits of observation, then galaxies about 10,000 million light years away will be receding at the speed of light and so will not be visible. This distance defines the boundary of the *observable universe*. It is hard to believe that light from these distant galaxies has been travelling for thousands of millions of years, and that we are seeing these galaxies not as they are now but as they were thousands of millions of years ago.

20

CHAPTER THREE

THE ASTRONOMER'S INSTRUMENTS

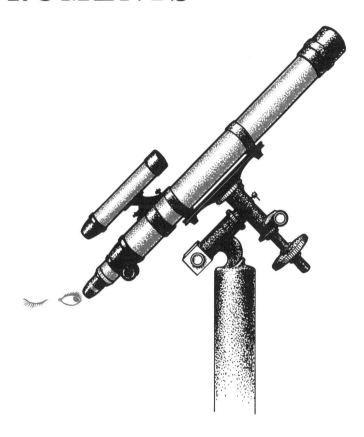

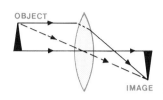

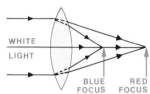

Formation of an image by a lens

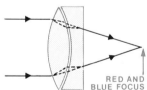

Chromatic aberration

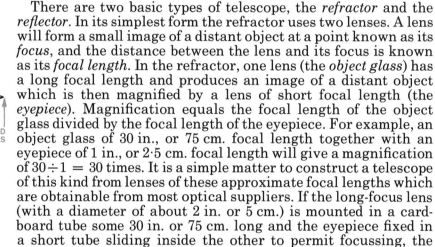

Achromatic lens

T he fundamental astronomical instrument is the telescope, and its purpose is twofold. Firstly it collects more light than the human eye and so allows the astronomer to see much fainter objects, and secondly it has a higher *resolving power*, i.e., it allows the observer to see finer details. Magnification, as such, is not a particularly important function of the telescope. An astronomer will use only a sufficiently high magnification to allow him comfortably to make out the details the telescope renders visible. This is determined by its resolving power and no amount of magnification will show details beyond this limit. In fact, excessively high magnification is a drawback in that the image tends to become blurred and faint.

There are two basic types of telescope, the *refractor* and the *reflector*. In its simplest form the refractor uses two lenses. A lens will form a small image of a distant object at a point known as its *focus*, and the distance between the lens and its focus is known as its *focal length*. In the refractor, one lens (the *object glass*) has a long focal length and produces an image of a distant object which is then magnified by a lens of short focal length (the *eyepiece*). Magnification equals the focal length of the object glass divided by the focal length of the eyepiece. For example, an object glass of 30 in., or 75 cm. focal length together with an eyepiece of 1 in., or 2·5 cm. focal length will give a magnification of $30 \div 1 = 30$ times. It is a simple matter to construct a telescope of this kind from lenses of these approximate focal lengths which are obtainable from most optical suppliers. If the long-focus lens (with a diameter of about 2 in. or 5 cm.) is mounted in a cardboard tube some 30 in. or 75 cm. long and the eyepiece fixed in a short tube sliding inside the other to permit focussing, the result will be a very satisfactory little telescope at a cost of less than a pound.

The drawback to a simple refractor is that simple lenses do not focus all colours to a single point. This defect, known as chromatic aberration, means that images are surrounded by fringes of false colour and resolution is impaired. It can largely be overcome by making the object glass from two or more components.

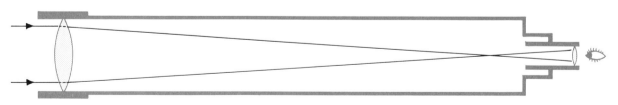

Principle of the refractor

The reflecting telescope uses a concave mirror to form an image and has the great advantage that all colours are reflected to the same focus so that chromatic aberration is eliminated. The first working reflector was built in 1672 by Isaac Newton and is illustrated below. Light from distant stars passes down a tube and is reflected from a concave (parabolic) mirror at the bottom to a focus which would lie in the middle of the tube. Clearly one cannot place an eyepiece there as the observer's head would get in the way of light proceeding towards the main mirror; Newton's solution was to place a small flat mirror inclined at 45° before the focus so as to reflect the focus point to the side of the tube where the eyepiece could be fixed.

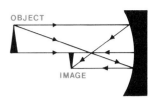

Formation of an image by a concave mirror

The Newtonian reflector is in common use, particularly among amateur astronomers because of its simplicity of construction and adjustment, but other types exist. In the *Cassegrain* reflector the secondary mirror is a convex one which reflects light coming from the main mirror (or primary) back down the tube and through a small hole in the primary to the eyepiece. Thus a long-focus telescope is compressed into a short length, and the observer looks in the same direction as the telescope is pointing. Like the refractor, these telescopes produce an inverted image of the object under scrutiny, but a variation on the Cassegrain system, the *Gregorian*, produces an erect image. Later, William Herschel used a simple design in which there were no secondary mirrors, but the primary was slightly tilted so as to bring the focus to the side of the tube at the top. This, however, led to other difficulties and the system is seldom seen now. More specialised reflectors have been designed for particular purposes, such as the Schmidt which allows large areas of sky to be photographed at once.

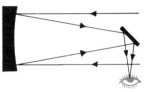

Principle of the Newtonian Reflector

As well as overcoming chromatic aberration, reflectors have other advantages over refractors. In particular they are cheaper to build and easier to mount. Consequently few large refractors are now built, and almost all new telescopes of any size are reflectors. The largest reflector at present has a 200-in. mirror whereas the largest refractor has a 40-in. diameter lens.

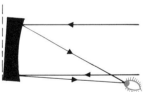

Principle of the Herschelian Reflector

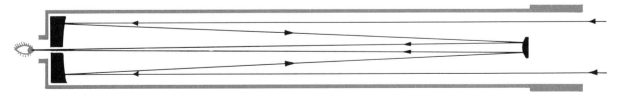

Principle of the Cassegrain Reflector

SPECTROSCOPES

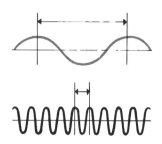

Wavelength: different colours of light have different wavelengths

Light is a form of *electromagnetic radiation* and can be regarded as a wave motion, analogous to water waves or sound waves except that light can travel through empty space and these other waves require water or air in which to travel. Like water waves we can think of light waves as having crests and troughs, where the distance between two successive crests is the *wavelength*. Different colours of light correspond to different wavelengths with red having long wavelength and blue short wavelength. These wavelengths are very small, being about 0·7 and 0·4 microns respectively (one micron = one millionth of a metre). When sunlight is passed through a glass prism it emerges as a rainbow band of colours (a *spectrum*). What actually happens is that a ray of light entering the prism is bent (*refracted*) away from its original direction. The shorter the wavelength the more it is refracted: thus blue is refracted more than red and the colours separated out.

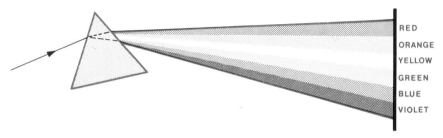

RED
ORANGE
YELLOW
GREEN
BLUE
VIOLET

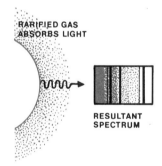

RARIFIED GAS ABSORBS LIGHT

RESULTANT SPECTRUM

Formation of dark lines in the sun's spectrum

When the spectrum of the sun is examined in detail, it is found to be crossed by dark lines called *absorption lines*. The hot surface of the sun emits light of all colours (a *continuous spectrum*), but before this light reaches us it passes through the cooler gas above the sun's surface. The atoms of this gas absorb light of particular wavelengths, giving rise to a series of absorptions which appear in the spectrum as dark lines. The elements present in the outer regions of the sun can be identified by their "fingerprint" patterns of dark lines. Likewise the spectra of stars can be similarly analysed. The instrument used to analyse spectra in detail is the spectroscope which takes light from a telescope and splits it up into component wavelengths by passage through a prism (or a device known as a *diffraction grating*), then focussing the spectrum onto a photographic plate, or to an eyepiece for direct viewing.

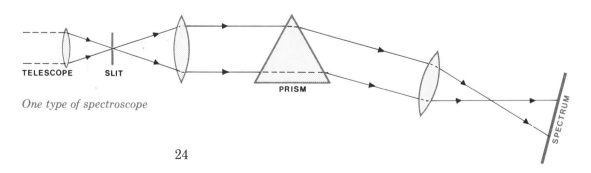

TELESCOPE   SLIT

PRISM

SPECTRUM

One type of spectroscope

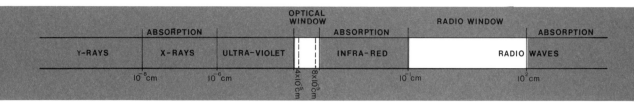

The electromagnetic spectrum: the waves usually studied with radio telescopes have wavelengths from a few millimetres to a few metres

A complete range of electromagnetic radiation exists, with wavelengths ranging from very much smaller than that of light to very much greater. The diagram above shows the different types of radiation which make up the *electromagnetic spectrum*. Analysis of radio and microwave radiation forms the basis of *radio astronomy*, which in the last two decades has emerged as a vital new tool of the astronomer. Many astronomical bodies, such as the sun, emit a certain proportion of radio waves, but others emit more radio than visible radiation, some of these being of great interest since they may be the most distant known objects in the universe. *Radio telescopes* can take various forms, but the simplest to understand is the dish-type which is very similar to the optical reflector, except that radio waves are focussed on to a receiver placed at the focus of the dish. It does not give a visible picture, but by scanning an object a contour map of its radio brightness can be built up.

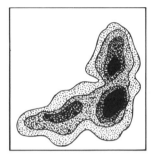

Contour map of a radio source

Radio telescope of the 'dish' type

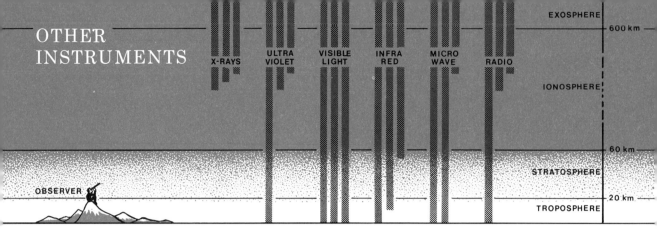

EXOSPHERE — 600 km —

X-RAYS · ULTRA VIOLET · VISIBLE LIGHT · INFRA RED · MICRO WAVE · RADIO

IONOSPHERE

— 60 km —

STRATOSPHERE

OBSERVER

— 20 km —

TROPOSPHERE

Transmission of the atmosphere for various kinds of radiation

Infra-red source: the quasar 3C273

X-ray source: the Crab Nebula

Artificial earth satellite

The modern astronomer has at his disposal a wide range of instruments, many of which would have been quite inconceivable a few decades ago. For example, the *photo-multiplier*, a device which converts the faint light from stars into electric currents, has enabled precise measurements of star brightness to be made, but really spectacular advances in astronomy are being achieved by the analysis of all kinds of electromagnetic radiation rather than just visible light and radio waves, and these developments are linked to space exploration and artificial satellites.

The earth-bound astronomer is severely hampered by our atmosphere which allows only certain types of radiation to reach ground level, namely visible and radio radiation and a small proportion of a type of radiation of longer wavelength than light known as *infra-red* (which we can detect as heat). Other wavelengths are absorbed or reflected away by the atmosphere. This is fortunate as far as the human race is concerned as the short-wave radiations—ultra violet, X-rays and Gamma rays—are extremely harmful to living tissue; but the astronomer could gain a great deal of information about the universe if he were able to collect and analyse these invisible radiations.

Our view of the universe is improved by setting up observatories at high altitudes on mountains. They are then above much of the dense, dusty lower reaches of the atmosphere and the quality of visible light is improved, while the proportion of ultra-violet and infra-red radiation received is greater than at ground level. If, however, we get above the atmosphere completely the entire electromagnetic spectrum is available. This is achieved either by using rockets fired to high altitudes carrying instruments which can be used for a few minutes before falling back to earth, or by flying experiments in artificial satellites which can stay in orbit round the earth for long periods of time. These approaches are proving particularly valuable in ultra-violet and X-ray astronomy.

26

CHAPTER FOUR

A CLOSER LOOK AT
THE SOLAR SYSTEM

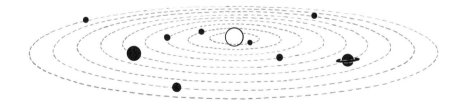

THE EARTH

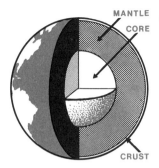

Interior structure of the earth

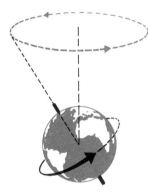

Precession

The earth is a solid body some 7,926 miles (12,800 km.) in diameter, rotating on its axis once every 23 hours 56 minutes, and moves around the sun in a nearly circular orbit of 93 million miles (150 million km.) radius. Its axis is not.perpendicular to the plane of its orbit, but is inclined at an angle of $23\frac{1}{2}$ degrees, so that sometimes the north pole and at other times the south pole is inclined towards the sun. This gives rise to our seasons. Of vital importance to human life is the atmosphere which is composed of 78% nitrogen and 20% oxygen together with small quantities of other gases such as carbon dioxide (which helps to retain heat that otherwise would radiate into space at night). The earth is unique in the solar system in having a surface nearly two thirds covered by water. Its internal structure has been determined largely by studying shock waves emanating from violent events such as earthquakes, and it appears that the earth has a central metallic *core* (which is at least partially molten, having a temperature of about 2000°C) with a radius of 2,100 miles (3,400 km.), above which lies the rocky *mantle*; on top of this is the *crust*, only some 30 miles (50 km.) thick. The earth's mean density is 5·5 gm. per cubic centimetre.

Although the moon has a mass of less than one eighteenth that of the earth, it does exert several effects upon the earth. It isn't quite true to say that the moon revolves round the earth; in fact both bodies revolve round their common centre of mass, a point known as the *barycentre*. Nonetheless, the barycentre does lie within the globe of the earth. The moon, together with the sun, exerts a pull on the oceans causing the tides, and furthermore, because the earth bulges very slightly at the equator the effect of the pull of these bodies is to make the axis of the earth wobble, or *precess*, so that the north celestial pole traces out a circle in the sky in about 25,800 years.

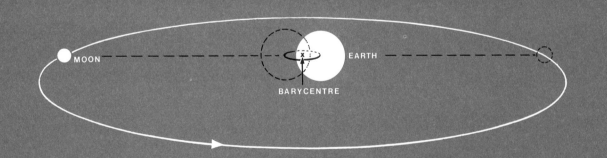

Motion of the earth and moon round the barycentre

MOON

EARTH

BARYCENTRE

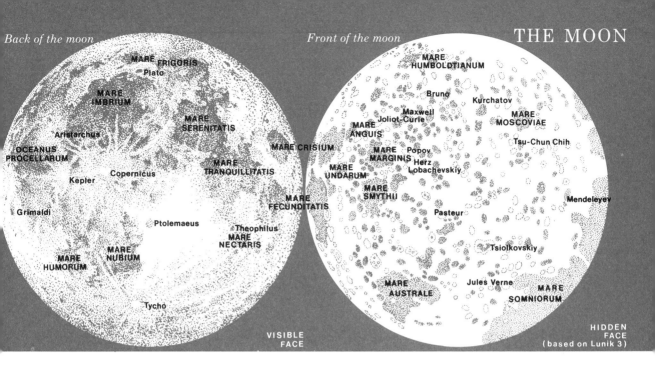

Back of the moon

Front of the moon

MARE FRIGORIS
Plato
MARE IMBRIUM
Aristarchus
OCEANUS PROCELLARUM
Kepler
Copernicus
Grimaldi
MARE SERENITATIS
MARE CRISIUM
MARE TRANQUILLITATIS
Ptolemaeus
MARE FECUNDITATIS
Theophilus
MARE NECTARIS
MARE HUMORUM
MARE NUBIUM
Tycho

MARE HUMBOLDTIANUM
Bruno
Kurchatov
Maxwell
Joliot-Curie
MARE ANGUIS
MARE MOSCOVIAE
Tsu-Chun Chih
MARE MARGINIS
Popov
Hertz
MARE UNDARUM
Lobachevskiy
MARE SMYTHII
Pasteur
Mendeleyev
Tsiolkovskiy
MARE AUSTRALE
Jules Verne
MARE SOMNIORUM

VISIBLE FACE

HIDDEN FACE
(based on Lunik 3)

The moon has a diameter of 2,160 miles (3,670 km.) and revolves round the earth in a period of 27·3 days. It also rotates on its own axis in the same length of time and consequently keeps the same side permanently turned towards us (it does *not* keep the same side turned towards the sun), this state of affairs probably being due to the strong gravitational effects caused in the moon by the earth. It is rather less dense than the earth (3·3 gm. per cubic centimetre), and because it's much less massive than the earth, the force of gravity at its surface is only one sixth of what we feel here on earth. The lower gravity means that astronauts can carry heavier objects and, should they fall over, they fall much more slowly than they would on earth.

The moon has no detectable atmosphere and suffers extremes of temperature ranging from over 100°C on the sunlit side to below −130°C on the night side. The main surface features are easily seen in small telescopes, these being the dark, fairly level plains, known as "seas" because the early observers thought they were full of water; the circular *craters* and the bright mountainous highland regions. There are other features, too: valleys, ridges and mountain chains, some of which are almost ten kilometres high. Analysis of the surface rocks by manned and unmanned exploration has shown that they are much older than terrestrial surface rocks, which seems to indicate that no really major surface changes have taken place for thousands of millions of years. Minor events, such as moonquakes, are still going on. Whether the craters and other features were due primarily to impacts of meteorites from space or to internal volcanic activity is not yet clearly settled.

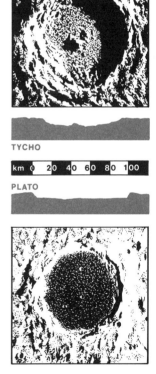

TYCHO

km 0 20 40 60 80 100

PLATO

Craters in plan and profile

29

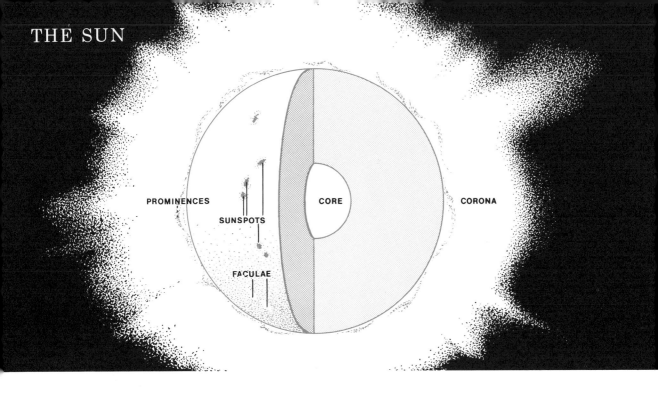

PROMINENCES

SUNSPOTS

FACULAE

CORE

CORONA

The corona seen during a total eclipse

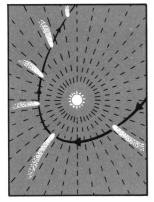

Effect of the solar wind on tail of a comet

The sun is a star, and as such is basically a huge ball of luminous gas, principally hydrogen. It has a diameter of 865,000 miles (1,390,000 km.), but although its volume is sufficiently great to contain well over a million earths its mass is only 330,000 times greater, thus its mean density is less than that of the earth. The sun rotates on its axis but because it's not a solid body it behaves in a peculiar way. At its equator it rotates in about 25 days, but further from the equator it takes longer, about 27 days at middle latitudes and even longer near the poles.

The visible surface is called the *photosphere* and has a temperature of 6000° Kelvin (the Kelvin, or Absolute, temperature scale begins at −273°C). The most prominent features are *sunspots*, regions whose temperature is between one and two thousand degrees lower than the rest of the photosphere. They appear dark, but only in contrast to their brighter surroundings, and their numbers vary considerably in a cycle of about 11 years. Above the photosphere lies the *chromosphere* (where spectral lines are produced) and the *corona*, the very tenuous solar "atmosphere" visible only during total eclipses. The sun emits a stream of charged particles, known as the *solar wind*, which stretches out beyond the earth. The sun is about 5,000 million years old and shines by converting hydrogen to helium by means of nuclear reactions in its core where the temperature exceeds 14,000,000°K. Four million tons of matter is converted into energy every second, but there is enough "fuel" for another 5,000 million years at the sun's present rate of consumption.

MERCURY is the nearest planet to the sun, though there have been suggestions that another may exist closer in. It is a small solid body with a diameter of about 3,000 miles (4,800 km.) and a density comparable to the earth, moving round the sun in a period of 88 days in an orbit markedly elliptical. Although its mean distance from the sun is 36 million miles (58 million km.) it ranges from 29 million to 43 million miles. It's an inhospitable world with no detectable atmosphere and surface temperatures ranging from about 400°C on the sunlit side to possibly −200°C on the dark side. For well over a century astronomers who studied its faint surface markings believed that Mercury rotated on its axis in 88 days and so kept the same face turned towards the sun, but in the last few years it has been established that it rotates in 59 days.

Mercury

Venus is very comparable in size to the earth, being 7,700 miles (12,200 km.) in diameter, and moves round the sun in a nearly circular orbit of radius 67 million miles (108 million km.), taking 225 days to do so. At its closest to the earth it comes nearer than any other planet, but until recently little was known about its surface because it is permanently covered in dense clouds. Since 1962, however, American and Russian space probes have investigated the planet in detail, and some of the Russian vehicles have landed. Their results show that Venus is an unpleasant place where the surface temperature exceeds 400°C, the atmospheric pressure is about 100 earth atmospheres and the atmosphere consists almost entirely of carbon dioxide. There is no prospect of our form of life existing there. Venus rotates on its axis in 243 days in the opposite direction to the other planets, and thus its sidereal "day" is longer than its year.

Venus' atmosphere may be so dense that light rays are bent round the planet

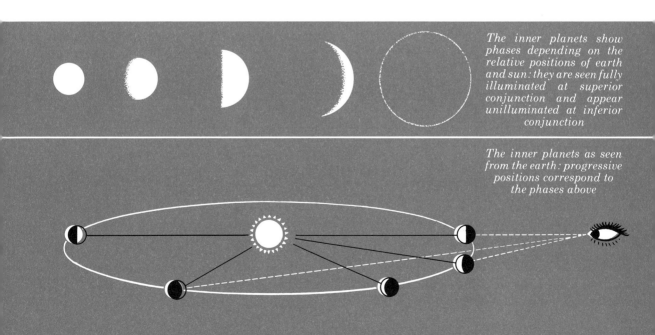

The inner planets show phases depending on the relative positions of earth and sun: they are seen fully illuminated at superior conjunction and appear unilluminated at inferior conjunction

The inner planets as seen from the earth: progressive positions correspond to the phases above

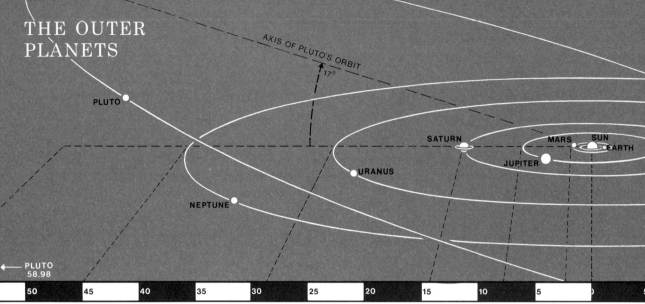

AXIS OF PLUTO'S ORBIT
17°

PLUTO

SATURN   MARS   SUN
EARTH

JUPITER

URANUS

NEPTUNE

← PLUTO
58.98

| 50 | 45 | 40 | 35 | 30 | 25 | 20 | 15 | 10 | 5 | 0 | 5 |

SCALE IN HUNDREDS OF MILLIONS OF KILOMETRES

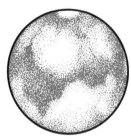

Three different aspects of Mars

MARS is in many ways the most interesting of the planets. A solid body 4,200 miles (6,800 km.) in diameter, it moves round the sun in an appreciably elliptical orbit at a mean distance of 142 million miles (228 million km.) with an orbital period of 687 days. At a favourable opposition it can approach to within 35 million miles of earth and can thus be studied in detail. Mars has a thin atmosphere in which clouds can sometimes be seen, some of which are known to be due to extensive dust storms. Telescopically, the planet appears reddish in colour with many pronounced dark markings which have been accurately mapped and which show seasonal variations. Like the earth it has white polar caps, at one time thought to be due to frozen water. The caps change in size with the seasons and may disappear completely in the Martian summer. It has two tiny moons, *Phobos* and *Diemos*, each of which is only about 10 miles (16 km.) in diameter, discovered in 1877, the same year as the famous Martian "canals" were first seen. These are apparently regular surface markings, once thought to be evidence of an advanced civilisation but now known to be natural features.

Since 1964, American and Russian space probes have had a closer look at Mars (a Russian vehicle landed in 1971, but ceased to transmit within minutes). They have shown that the atmosphere is about as thin as that of the earth at a height of 25 miles (40 km.) and consists primarily of carbon dioxide with negligible water vapour. The polar caps seem to consist of frozen carbon dioxide. Some features do look remarkably like the results of water erosion and one theory is that a rainstorm occurs once every 25,000 years. Temperatures range between 20°C and below −70°C, which makes it all rather unfriendly.

JUPITER is by far the largest planet in the solar system, being 318 times more massive than the earth and having a diameter at its equator of 88,700 miles (145,000 km.). It lies at a mean distance from the sun of 483 million miles (778 million km.) and takes 11·86 years to complete an orbit. It does not seem to have a solid surface and, as its low density (1·3 gm. per cubic centimetre) suggests, is primarily a gaseous body. It takes only 9 hours 51 minutes to rotate on its axis and consequently bulges at the equator to such an extent that the diameter from pole to pole is about 5,000 miles less than at the equator. Telescopically, its disc is seen to be covered with cloud belts, the principal constituents of which are hydrogen compounds, such as methane and ammonia, and whose temperature is about −130°C. But density and temperature increase rapidly towards the centre of the planet where there may be a dense metallic core perhaps ten times as massive as the earth. Jupiter has twelve moons, four of which are comparable with our moon and easily seen with small telescopes.

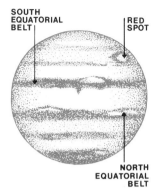

The cloud belts of Jupiter

Saturn is another giant planet, similar in many ways to Jupiter but almost twice as far from the sun at a mean distance of 886 million miles (1,420 million km.). It takes 29·5 years to complete an orbit. Its composition is similar to Jupiter but its density is even less: at 0·7 gm. per cubic centimetre it is less dense on average than water. Even so it is 95 times as massive as the earth and has an equatorial diameter of 75,000 miles (121,000 km.). As a consequence of its rapid rotation rate (10¼ hours) it is more markedly flattened than Jupiter. Saturn is unique in having a system of rings composed of swarms of tiny particles (mainly lumps of ice) revolving round about it, and although the system is 175,000 miles (280,000 km.) across, it is only about 5 or 10 miles thick. Saturn has ten moons, one of which—Titan—is larger than Mercury.

Varying aspects of Saturn's rings

Saturn and its ring system

SHADOW OF PLANET ON RINGS

SOUTH EQUATORIAL BELT

RING A
CASSINI DIVISION
RING B
RING C

THE OUTERMOST PLANETS

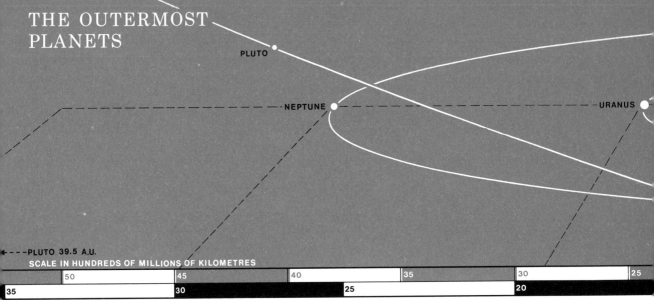

PLUTO

NEPTUNE

URANUS

PLUTO 39.5 A.U.

SCALE IN HUNDREDS OF MILLIONS OF KILOMETRES

| 50 | 45 | 40 | 35 | 30 | 25 |

| 35 | 30 | 25 | 20 |

SCALE IN ASTRONOMICAL UNITS

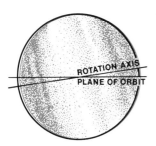

Uranus

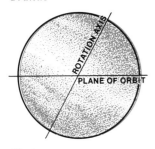

Neptune

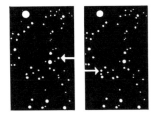

Motion of Pluto in three days

U*RANUS*, discovered telescopically in 1781 by William Herschel, is a giant planet similar in general constitution to Jupiter but with a temperature of $-190°$C. Its mean distance from the sun is 1,780 million miles (2,870 million km.) and it takes 84 years to complete an orbit. Its equatorial diameter is 29,600 miles (49,000 km.) and it rotates in 10 hours 48 minutes. A curious feature is that its axis is inclined at 98 degrees, so it sometimes moves along pole first. Uranus has five moons, all of them smaller than ours.

Neptune is also a giant planet and was discovered in 1846 very close to the position predicted by the mathematician J. C. Adams and U. le Verrier. Uranus has been observed to move somewhat erratically, and it materialised that this was due to the gravitational attraction of the at that time unknown Neptune. Similar in size to Uranus, it has two moons, lies at a mean distance from the sun of 2,790 million miles (4,500 million km.) and takes 164·8 years to complete an orbit, while rotating on its own axis in 14 hours.

Pluto is the most distant planet known in the solar system, but for part of the time it actually comes closer in than Neptune because it has a highly eccentric orbit. This is the situation at present, although Pluto will not be at its closest until 1989. Its mean distance from the sun is 3,600 million miles (5,900 million km.) and its orbital period 248 years. Discovered in 1930, its existence was predicted in the U.S.A. in 1914 by Percival Lowell after his analysis of the motions of Neptune and Uranus; but Pluto is much smaller than expected, having a diameter of only about 3,600 miles. It may therefore be that there is a more distant planet still to be found, and attempts are being made to pin it down.

34

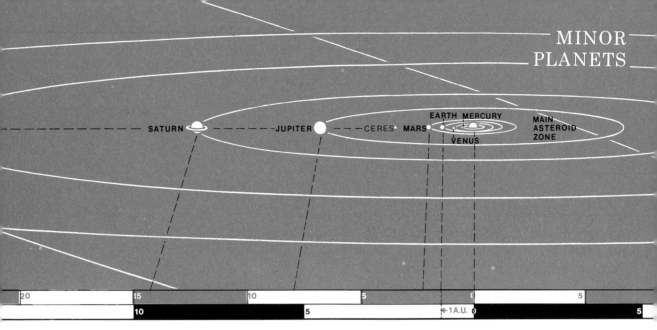

SATURN — JUPITER — CERES MARS — EARTH MERCURY — VENUS — MAIN ASTEROID ZONE

Besides the nine major planets, there is a host of smaller objects, known as *minor planets*, or *asteroids*. The story of their discovery is of some interest. In 1772 Johan Bode drew attention to a curious numerical relationship between the distances of the planets from the sun, which came to be known as *Bode's Law*. As shown on the right, if we take the series 0, 3, 6, etc., where each successive number is double the preceding one, add 4 to each and divide by 10, we arrive at the series, 0·4, 0·7, 1·0 and so on. The distances of the planets known in Bode's time (Mercury to Saturn) when measured in astronomical units agreed very well with the series. We now know that Uranus, at least, also agrees with Bode's Law. It was apparent to Bode's contemporaries that no known planet existed at the distance 2·8 A.U., and a search was begun for this "missing planet".

In 1801 a planet was found orbiting at a distance of about 2·7 astronomical units as expected. Named *Ceres*, it proved to be a tiny world only 450 miles (700 km.) in diameter with an orbital period of 4·6 years. Within a few years several more had been found, and now some two thousand have had orbits calculated, and it's estimated that at least 40,000 exist. Ceres remains the largest, and there are only about a dozen with diameters exceeding 120 miles (200 km.), the rest ranging in size down to a few tens of metres. Most of them are confined to the region between Mars and Jupiter, but some venture closer in; *Icarus*, for example, can approach to within 20 million miles (32 million km.) of the sun, while others, such as *Hermes*, can come within a few hundred thousand miles of earth. Their paths do not intersect so there's no danger of collision. The origin of the minor planets remains a mystery: they may be the remnants of a shattered planet, or alternatively matter which did not form into planets in the first place.

BODE'S LAW

PLANET	DISTANCE FROM SUN (A.U.)	BODE'S LAW DISTANCE
Mercury	0·39	0·4
Venus	0·72	0·7
Earth	1·00	1·0
Mars	1·52	1·6
?	–	2·8
Jupiter	5·21	5·2
Saturn	9·54	10·0
Uranus	19·18	19·6
Neptune	30·1	
Pluto	39·5	38·8

35

COMETS AND
METEORS

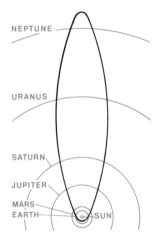

Orbit of a comet

Meteor radiant

COMETS can be among the most impressive astronomical phenomena, a naked-eye comet having a bright head and a long, tenuous, luminous tail stretching away from it. Towards the end of the nineteenth century several bright comets were seen some of which had tails stretching right across the sky. Astronomically, comets are insignificant objects, usually moving in very elongated elliptical orbits round the sun. The best-known comet is *Halley's*, (named after Edmond Halley who first showed that comets move in this way) which orbits the sun in 76 years and is due back again in 1986. Bright comets are rare, generally having long orbital periods, possibly millions of years. A comet has a head, or *nucleus*, composed of a loose aggregation of gas, dust and solid particles. The *tail*, which may be millions of kilometres long, is due to tenuous dust and gas ejected from the nucleus by radiation pressure and the solar wind as it approaches the sun. It always points away from the sun no matter what direction the comet is going.

Meteors, or "shooting stars", are tiny particles, usually no bigger than grains of sand, which enter our atmosphere at speeds up to 45 miles (70 km.) per second and burn up as a result of friction. Isolated, or *sporadic*, meteors are seen, but meteor *showers* also exist, being due to material scattered along certain orbits in space, some of which are associated with former comets. As the earth crosses these orbits, meteors appear to come from certain points in the sky called *radiants*. *Meteorites* are much larger rocky or metallic objects which can survive passage through the atmosphere and may reach ground level. They are perhaps related to the minor planets, but like comets and meteors are no more than space debris.

CHAPTER FIVE
A CLOSER LOOK AT THE STARS

PLEIADES

CONSTELLATIONS

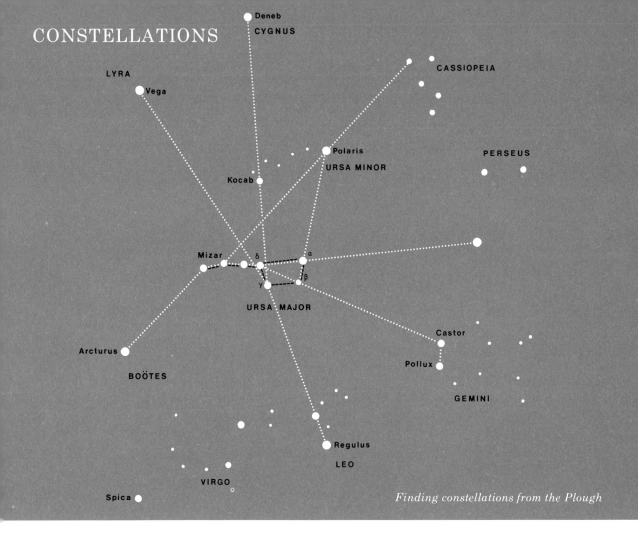

Finding constellations from the Plough

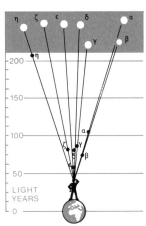

Relative distances of the stars in the Plough

At first glance the sky appears to be a confusing mass of stars, but a closer inspection will show certain patterns begin to emerge. These patterns, or groupings, of stars are known as *constellations*, 88 of which have been identified on the celestial sphere. Ancient astronomers named the constellations after mythological figures, and although the shapes of some do bear a certain relationship to their subjects, a highly vivid imagination is essential to identify others! Constellations have no physical significance: generally the stars contained within them are in no way related to each other, and the shape a constellation would have when seen from elsewhere in space will be quite different from the shape we see. The simplest way to get to know the constellations is to identify a few of the most obvious ones and use these as guidelines for the others. In the northern hemisphere, the best-known constellation is the Great Bear (Ursa Major), often called the *Plough* because of the shape made up by its seven brightest stars.

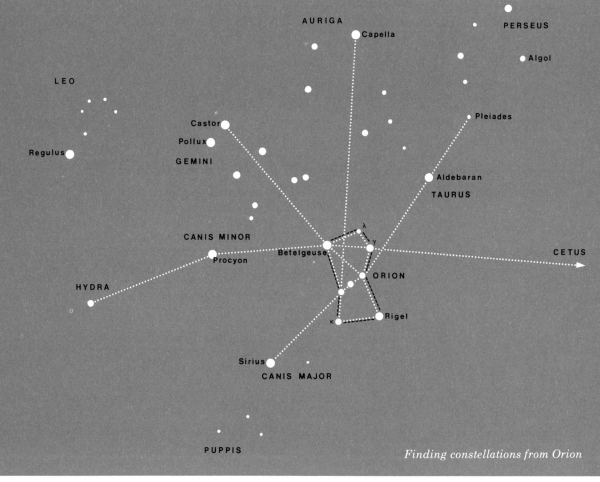

AURIGA
Capella
PERSEUS
Algol
LEO
Pleiades
Castor
Pollux
GEMINI
Regulus
Aldebaran
TAURUS
λ
γ
CANIS MINOR
Betelgeuse
ORION
CETUS
Procyon
HYDRA
Rigel
κ
Sirius
CANIS MAJOR
PUPPIS

Finding constellations from Orion

The two stars at the front of the blade of the Plough are called the Pointers, as a line through them eventually leads to the isolated star *Polaris*, known as the Pole Star, as it lies only one degree from the north celestial pole. As shown on the facing page, the Plough can be used to locate several prominent constellations, including *Cassiopeia* (The Lady in the Chair), *Auriga* (the Charioteer), *Gemini* (the Twins), *Leo* (the Lion), *Bootes* (the Herdsman) and *Cygnus* (the Swan). It is, too, a circumpolar constellation at latitudes north of 40° North. *Orion* (the hunter) is quite unmistakable and has the great advantage that it straddles the celestial equator and can thus be seen from almost anywhere on earth. It contains two of the brightest stars in the sky, the red Betelgeuse and the blue-white Rigel. The three stars in the middle form Orion's "belt", and a line taken downwards through them leads to *Sirius* (in Canis Major), the brightest star in the sky. In the opposite direction the line leads to *Taurus* (the Bull).

○ SUN

EDGE OF
BETELGEUSE

The star Betelgeuse in Orion compared to the sun

39

SEASONAL
STAR CHARTS

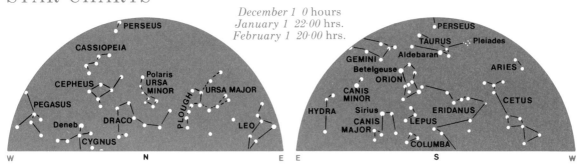

The stars which can be seen at any time depend on the season of the year, and on the latitude of the observer of the surface of the earth. The maps here have been drawn for latitude 45°N but will be quite useful between latitudes 30°N and 60°N. Observers located further south will see more of the southern constellations and less of the northern ones, and vice-versa. The maps themselves are in pairs, one showing the sky as seen when facing south, the other showing the sky as seen when facing north. Pairs of maps are shown for each season, but the constellations rise about two hours earlier each month; the times at which the maps are exactly right change from month to month, and are depicted here. As time is measured by the rotation of the earth, these figures, local time, apply in all parts of our planet

The winter sky is in many ways the most spectacular. Looking south, the most prominent constellation is Orion which can be used to locate the others. Upwards and to the right is the bright red star Aldebaran in the constellation Taurus, and a little further on is a beautiful cluster of stars, the Pleiades (or "Seven Sisters"). Above and to the left of Orion is Gemini with the two bright stars Castor and Pollux. Below Gemini lies Procyon, the bright star in Canis Minor, and lower down again is Sirius, the brightest star in the sky. Leo is rising in the east. Facing north, the Plough seems to be standing upright and as already noted can be used to locate the pole star and the "W" shape of Cassiopeia, through which runs the Milky Way. Pegasus and Andromeda (in which the famous galaxy may be seen) are in the west, and the bright star vertically overhead is Capella, in Auriga.

Looking south in spring, Leo is high in the sky while Orion is low in the west. West of Leo is the faint constellation Cancer (the Crab), in the centre of which is a beautiful cluster of stars—the Beehive—well seen in binoculars. Virgo (the Virgin) is in the south-east, and the bright star Arcturus, in Bootes, is almost due east. In the north, the Plough is now high in the sky, upside-down in appearance. Cassiopeia, on the opposite side of the Pole Star is low down. Low in the north-east is the bright Vega (in Lyra) and further east is Hercules. West of north are Taurus, Auriga, Perseus, and higher up, Gemini.

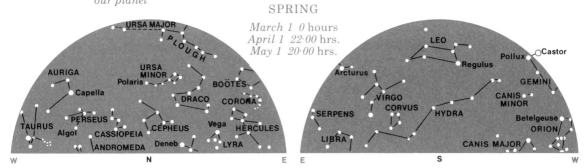

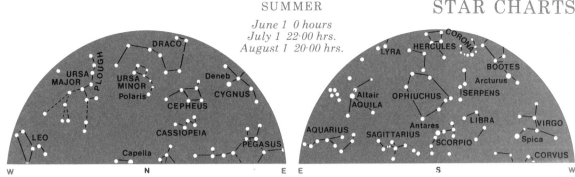

The summer sky is dominated by the "Summer Triangle" which is high in the south and made up of the three stars Deneb, in the constellation of Cygnus; Altair, in Aquila (the Eagle); and Vega, which is now almost vertically overhead. The Milky Way passes through these three constellations. West of Lyra lies Hercules, and below is Ophiuchus, a rather vague constellation. Near the horizon is the bright red star Antares, in Scorpio (the Scorpion), while high in the south-west is Arcturus, above which is the beautiful little constellation Corona Borealis. In the northern sky, the Plough is pointing downwards, and Capella is low on the horizon. Andromeda and Pegasus are rising in the north-east, while Cepheus lies between Cassiopeia and Cygnus.

Looking southwards in autumn, the four stars which make up the distinctive "square" of Pegasus are high in the sky with Andromeda above and to the left. Eastwards are the faint constellations of Aries (the Ram), Pisces (the Fishes) and Cetus (the Whale). Below Pegasus and due south is Aquarius, and lower down is the bright red star Fomalhaut in Piscis Austrinis. The summer triangle is still high in the south-west, while between Aquila and Cygnus and a little to the east lies the tiny constellation Delphinus. Looking north, the Plough is the correct way up and quite low. The Pole star forms the end of the tail of the much fainter Little Bear. Capella is in the north-east, above which is Perseus, while Aldebaran is rising almost due east, and Orion will follow.

AUTUMN

September 1 0 hours
October 1 22·00 hrs.
November 1 20·00 hrs.

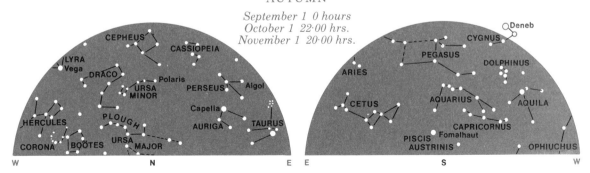

BRIGHTNESS
AND COLOUR

Star B appears brighter than star A although A is actually more luminous than B

The apparent brightness of a star as we see it from the earth is its *apparent magnitude* (m). A fairly bright star such as Aldebaran is of First Magnitude (i.e., has m=1), a rather fainter one such as the Pole Star has m=2, and so on. The faintest star visible to the naked eye has m=6, and is 100 times fainter than a star of m=1. The faintest detectable stars have m=23. At the other end of the scale, stars brighter than first magnitude can have m=0 (e.g. Alpha Centauri) or even minus values, such as Sirius with m=−1·5. The sun has m=−26. A star's apparent magnitude depends upon both its real brightness and its distance from us. It can happen that a star which is really very luminous can appear faint simply because it's far away. To compare the real brightnesses of stars, astronomers use *absolute magnitude* (M), this being the apparent magnitude a star would have if placed at a standard distance from the earth of 32·6 light years. The sun has M=4·8 while Aldebaran has M=−0·1. Aldebaran is thus inherently nearly 100 times more luminous than the sun.

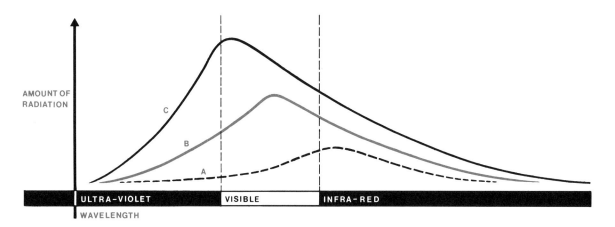

COLOUR AND TEMPERATURE

The graph shows the amount of radiation of different wavelengths emitted by stars of different temperatures: A (3,000°K), B (6,000°K) and C (11,000°K)

The different colours of the stars indicate their different surface temperatures. Red stars are relatively cool (about 3,000°K), yellow stars such as the sun are warmer (about 6000°K), white stars such as Sirius are hotter (11,000°K), and blue stars are still hotter, with temperatures up to about 30,000°K. Stars give out all kinds of radiation, but with increasing temperature they emit more and more short-wave radiation. Since blue light has shorter wavelength than red, the hotter they are the bluer they seem.

The basic technique for measuring stellar distance is the method of trigonometrical *parallax*. In principle the method is simple: if we observe the position of a nearby star (X) against the background of distant stars when the earth is at E, and again six months later when the earth is on the opposite side of the sun (S) at point F, then because the two points E and F are separated by the diameter of the earth's orbit, the star will appear to be displaced in position. If we measure the angle EXS and apply simple trigonometry to the triangle SXE then we can find the distance of the star. The angle EXS is called the *annual parallax*. If a star had a parallax of exactly one second of arc it would lie at a distance of one *parsec*, a unit frequently used and equalling 3·26 light years; thus the standard distance for absolute magnitude measurements, 32·6 light years, equals 10 parsecs.

Annual parallax

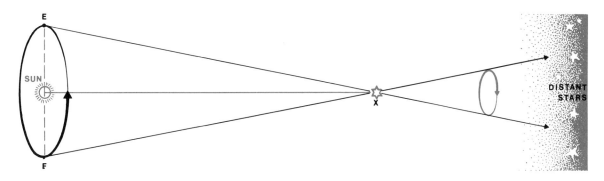

Unfortunately, even the nearest star has a parallax of only 0·76 seconds of arc, and because of the difficulty of making such tiny measurements, simple trigonometrical parallax cannot be used for stars much more than 50 parsecs away. Other methods of parallax which, for example, can be applied to clusters of stars, can be used out to greater distances, but most methods of obtaining large distances rely on a different approach. By studying the spectra of stars, or certain other properties, it is possible to work out what their absolute magnitudes ought to be, and it's then a simple matter to compare these with their observed apparent magnitudes and obtain their distances.

No star is near enough for it to be seen as a disc which could be measured (even the largest telescope will show a star only as a point of light), and although indirect observational methods are possible only a few stars have had diameters measured by these means. There is another method however: if we combine a star's temperature, obtainable from its colour, with its luminosity, derived from its absolute magnitude, we get the surface area, and since stars are spherical we can calculate their diameters too.

Three related quantities for a star: knowledge of two of these will usually enable the other to be calculated

43

VARIABLE STARS

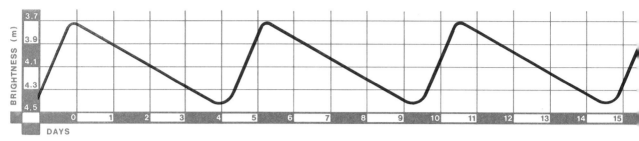

Light curve, showing variations in brightness of Delta Cephei

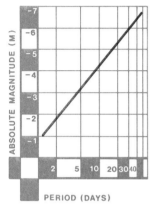

The period luminosity law for cepheids

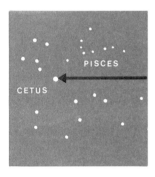

Location of Mira Ceti

Variable stars, are stars which vary in brightness, some in a regular, periodic fashion, others in quite unpredictable ways. Of great importance are the *Cepheid* variables. These are stars which vary with regular periods ranging from a few days to a few tens of days, and are named after the first star of this type to be studied—Delta Cephei (i.e., Delta in the constellation Cepheus) which varies between magnitude 3·7 and 4·4 in a period of 5·3 days. What is important about the Cepheids is that there is a direct relationship between the periods of variation and the absolute magnitudes of these stars: the longer the period, the greater the luminosity (which corresponds to lower absolute magnitude). This relationship is the *Period–Luminosity Law*, and is of fundamental importance since it means that be simply measuring the period of a Cepheid its absolute magnitude and distance can be found by comparing m and M. Since Cepheids are bright stars, they can be used as "standard candles" for distance measurements even as far as some of the nearer galaxies.

There are also *short-period variables*, such as the RR Lyrae stars which have periods of only a few hours, and at the other end of the scale *long-period variables* which may take hundreds of days to complete a cycle. They are not as regular as Cepheids in their behaviour, and there is no relationship between period and luminosity. The best-known example of a long-period variable, Mira Ceti, is clearly visible to the naked eye at its maximum (m = 2), but disappears from view at minimum when its apparent magnitude drops to 9. Like other long-period variables, it is a giant red star. There are many types of *irregular variables*, some of which flare up briefly from their normal brightness and others which, without warning, drop down temporarily. The most drastic "variables" are exploding stars—*novae* and *supernovae*. Novae are stars which shed their surface layers in an outburst that causes them to flare up in brightness by a factor of a thousand or more before fading back to normal. Supernovae occur when stars suffer collossal internal explosions, shedding a large proportion of their material into space and sometimes shining for a few days as bright as an entire galaxy.

44

If two stars lie in the same direction, but are in no way connected, they may look in the telescope like a double star. Such pairs of stars, aligned by chance, form *optical doubles*.

At the same time there are many double stars physically related to each other and revolving around their common centre of mass. Such stars are called *binaries*.

Some binaries can be easily studied, because the two components are well separated. They will have long orbital periods, probably hundreds of years. Some are visible to the naked eye, such as the star Mizar, the middle star of the handle of the Plough, and its faint companion Alcor. Sometimes, though, the companion star is so faint it cannot be seen, in which case its presence may be deduced by the way its gravitational pull causes the primary star to wobble as it moves through space. Sometimes, too, binaries are so close that they are seen as one star, but analysis of the spectrum of such "stars" will often show their binary nature. Multiple stars with three, four or more components exist, such as the "double-double" in Lyra. If the orbit of a binary is seen edge-on from the earth, then one star can pass in front of the other, eclipsing it and causing the combined brightness to fall. Such stars are *eclipsing binaries*. If one star is fainter than the other the fall in brightness may be less when it passes in front of the other than when the rôles are reversed. One eclipsing binary visible to the naked eye is Algol in Perseus.

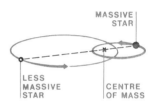

Orbits of binary stars round centre of mass

Eclipsing binary: the light curve relates to the orbit diagram (right)

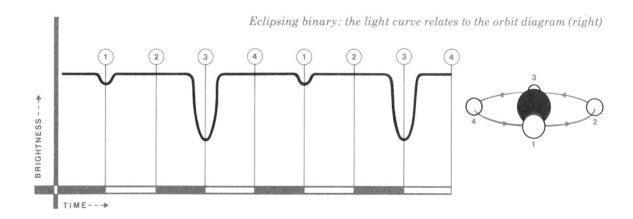

LIFE CYCLE
OF A STAR

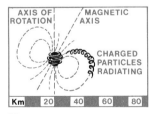

GAS CLOUD

CONTRACTING

Evolution of a star like the sun

MAIN SEQUENCE
STAGE

RED GIANT

WHITE
DWARF

DARK
BODY

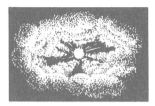

Nova

Supernova remnant

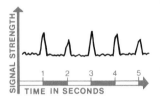

SIGNAL STRENGTH

1 2 3 4 5
TIME IN SECONDS

Radio signal from a pulsar; pulsars have been identified with neutron stars

AXIS OF
ROTATION

MAGNETIC
AXIS

CHARGED
PARTICLES
RADIATING

Km 20 40 60 80

Neutron star

Stars such as the sun exist for thousands of millions of years and clearly astronomers can never watch a particular star through its life cycle. However, the stars we see have different ages, and by observing many bodies at different stages of evolution astronomers can piece together the general way in which they are born, live, and finally die. A star like the sun begins as a cloud of gas which contracts under its own gravitational forces, and as it does so it becomes steadily hotter. It continues to contract until the temperature and pressure near its centre are sufficiently high for nuclear reaction, similar to that which take place in the hydrogen bomb, to begin to convert hydrogen into helium. The energy from such reaction prevents the star contracting any further and will continue to keep the star shining for, in the case of the sun, 10,000 million years. While at this stage it is said to be a *main sequence* star.

Eventually, though, the hydrogen "fuel" in the centre becomes exhausted and the next stage begins. Surprisingly, the star begins to expand and becomes much brighter, evolving into what is called a *red giant* (such as Aldebaran), a star up to hundreds of times larger than our sun. It does not remain a red giant for long, and soon, when it can no longer produce any nuclear energy, it collapses on itself and becomes a very dense *white dwarf*. Such stars may be smaller than the earth and can be so dense that a bucketful of white dwarf material would weigh hundreds of tons! Eventually, white dwarfs fade away, and the star ends as a cold, non-luminous body.

This is a very simplified account. At some stage, for example, the star may become unstable and then pulsate (as do the Cepheid variables) or shed material in a nova explosion. Stars which are much more massive than the sun evolve more quickly because, being more luminous than the sun, they use up fuel at a much higher rate and may, in fact, become supernovae. The remnants of these stars sometimes collapse into incredibly dense objects called *neutron stars*, which are millions of times denser than even white dwarfs. Finally, really massive stars may become *black holes*, objects so dense that gravity prevents light escaping from them and they become invisible.

46

CHAPTER SIX

A CLOSER LOOK AT THE GALAXIES

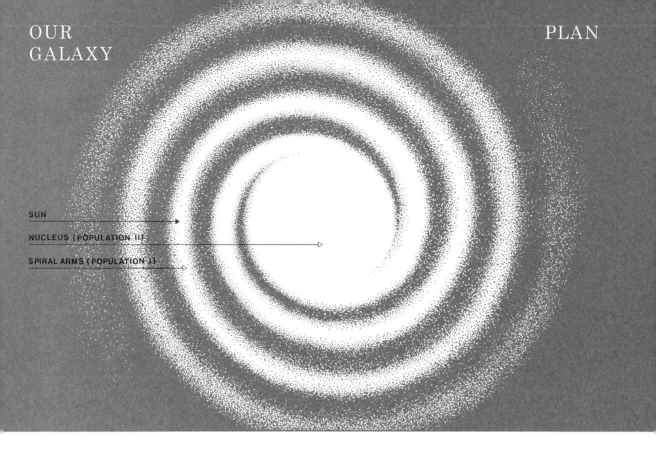

SUN

NUCLEUS (POPULATION II)

SPIRAL ARMS (POPULATION I)

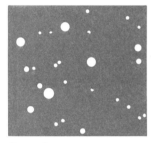

Open cluster

Globular cluster

Our galaxy contains an estimated hundred thousand million stars as well as appreciable quantities of matter in the form of gas and dust. In general shape it consists of a central nucleus some 20,000 light years thick surrounded by a disc of stars and other material 100,000 light years in diameter and 1,000 light years thick. The sun is located in the disc about 30,000 light years from the centre, which from our point of view lies in the direction of the constellation Sagittarius. The stars in the central nucleus are predominantly old red ones (called Population II) and there is little hydrogen gas located there. The disc contains most of the gas and dust as well as a high proportion of younger stars (Population I), many of which are second generation stars having formed from hydrogen gas into which has been mixed the material ejected from earlier supernovae explosions.

The galaxy also contains clusters of stars, of which there are two basic types—*open clusters* and *globular clusters*. Open clusters contain Population I stars and are usually fairly young on the cosmic time-scale. They contain up to a few hundred stars and often appreciable gas and dust. Globular clusters are compact aggregations of up to hundreds of thousands of old red stars of Population II and contain little gas or dust.

48

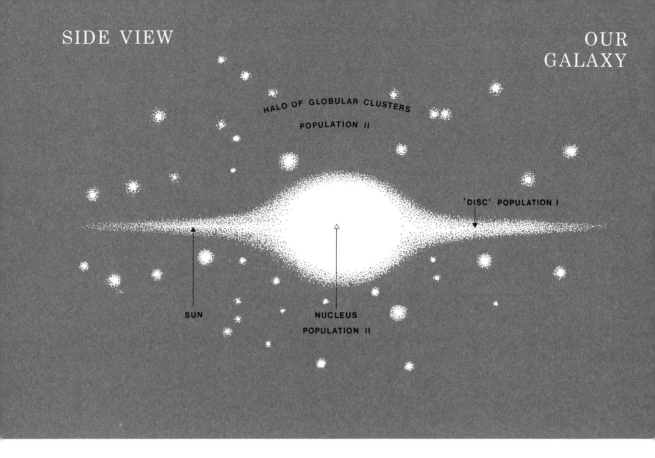

The vast quantity of gas spread through the disc of the galaxy has the unfortunate effect of obscuring the centre of the system from our view with optical telescopes, so it's extremely difficult to determine the structure of our galaxy. We can, however, make use of the globular clusters. There are about a hundred of these and they lie well clear of the disc of the galaxy (and so are not obscured by dust) in a sort of halo. If we assume that the centre of the system of these clusters coincides with the centre of the galaxy as a whole, then since their positions can be measured an estimate of the distance to the centre of the galaxy can be made. This agrees well with distance measured in other ways. Hydrogen gas in the galaxy emits radio waves of wavelength 21 cm. detectable by radio telescopes and, since this radiation isn't affected by the dust, the distribution of hydrogen gas in the system has been accurately mapped to give a detailed picture of the galaxy's structure. The disc of the galaxy has a spiral structure, with "arms" of gas and stars spreading out from the nucleus. The sun is located near the edge of one of these arms. Recent measurements have shown that the galaxy centre is a strong source of radio and infra-red radiation, and it looks as though large quantities of energy are being produced there. The galaxy centre may, in fact, be quite a violent place.

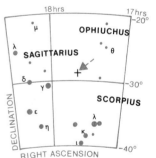

Direction of galaxy centre

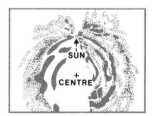

*Radio map of hydrogen gas
in our galaxy*

49

NEBULAE

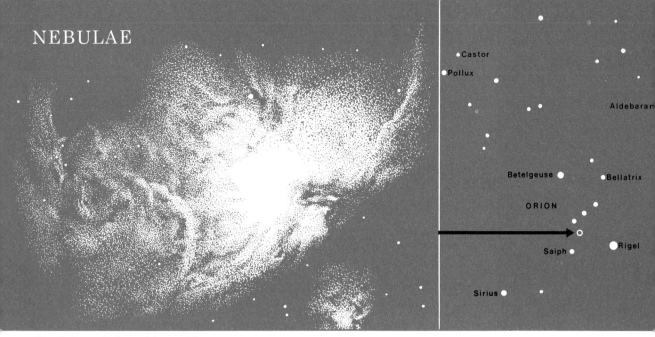

The Orion nebula and its position

The Horse-Head nebula

Origin of dark nebulae

With large telescopes we find many faint luminous patches known as *nebulae* (from the Latin for "clouds"). Some of these turn out to be galaxies in their own right, lying beyond our own system, but others are clouds of glowing gas, predominantly hydrogen, within our own galaxy. Although it's sometimes hard to distinguish between the two visually—for with distant galaxies we can't see the individual stars—analysis of their light with the spectroscope makes the difference clear. Galaxies have continuous spectra with absorption lines as a result of the combine light of thousands of millions of stars, whereas true nebulae show only bright emission lines corresponding to the atoms making up these gas clouds. The clouds are extremely tenuous, having only a few hundred atoms per cubic centimetre, and they shine by absorbing light from hot stars embedded within them and re-emitting it. Such nebulae are called *emission nebulae*, a good example of which is the Great Nebula in Orion, a cloud some 30 light years in diameter.

There are also luminous *reflection nebulae* and *dark nebulae*, both of which are due to the solid particles which make up the interstellar dust. A reflection nebula arises from light from a star being reflected from neighbouring dust particles, while dark nebulae are dark patches in the sky caused by dust clouds obscuring light from the stars behind. Orion also contains a fine example of a dark nebula (called the "Horse-Head"). Interstellar material (in the proportion 99% gas to 1% dust) composes between 10 and 20 per cent of the mass of our galaxy, and all the nebulae consist of a mixture of gas and dust. But emission nebulae arise only when certain types of hot stars cause the gas to shine, while the dust causes dark nebulae.

50

Elliptical

There are various types of galaxies in the universe, the main types being elliptical, spiral or irregular in shape. *Elliptical* galaxies range from spherical objects to rather flattened systems whose thickness may be only one seventh of their breadth. Usually they contain Population II stars and little, if any, gas and dust—possibly they are old systems. Some are huge, many times more massive than our galaxy, and these may be strong radio sources, but others are tiny systems (*dwarfs*).

Spiral

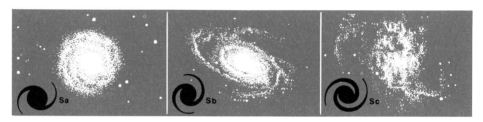

Spiral galaxies, of which our system is one, are of two basic types: ordinary spirals, in which arms of stars and interstellar matter spiral out from the central nucleus, and *barred spirals*, where the arms emanate from the ends of a luminous "bar" of material straddling the nucleus—the origin of which is unknown. Spirals are classified according to how tightly the arms are wound (from Sa to Sc) and contain many Population I stars and considerable gas and dust. Some are radio sources.

Irregular

Irregular galaxies have no orderly structure and are often relatively small systems. Our own galaxy is accompanied by two small irregulars—the Magellanic Clouds—which can be seen with the naked eye in the southern hemisphere. Such systems may or may not contain much gas. The irregular (M82) is a strong radio source, and appears to be exploding.

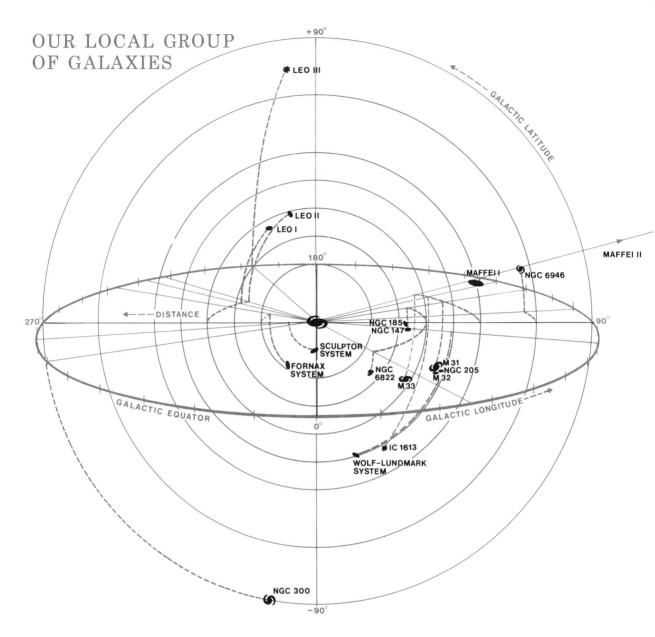

Our galaxy is a member of a small cluster of galaxies known as the *Local Group*. The exact number of galaxies in the group is uncertain, as some are rather faint systems, but there are at least twenty, and there may be as many as thirty. Most of these are small ellipticals or irregulars, but there are certainly four spirals—our galaxy; the Andromeda, similar to our own; a smaller spiral, named M33; and Maffei II, a galaxy hidden from view by dust until discovered in 1971 by infra-red astronomers. The elliptical, Maffei I, was found similarly. The group is 4,000,000 light years across.

The distance of an external galaxy was first measured in the U.S.A. in 1923 by Edwin Hubble. He detected bright Cepheid variables in the Andromeda Galaxy and used the Period–Luminosity Law to determine distance. Galaxies up to ten times further away can be measured in this way, but beyond that different methods are needed. For example, the absolute magnitudes of bright super-giant stars, globular clusters or even supernovae may be related to their apparent magnitudes to give distances.

Almost without exception it's found that every galaxy beyond the boundaries of the Local Group is moving away from us. This has been determined by means of the *Doppler Effect*. When a source of light moves away from us, its wavelength increases as though the waves were being "stretched". The change in wavelength is proportional to the velocity at which the source is receding and, since red corresponds to long wave light, the effect is often called the *red-shift*. In the spectrum of a receding galaxy the spectral lines will be displaced towards the red end of the spectrum and by measuring this displacement the galaxy's velocity is found.

Hubble was able to show that the greater the distance of a galaxy the faster it was receding, and that this velocity was related to distance by *Hubble's Constant*, H, which has a value between 75 and 100 km. per sec. per megaparsec (i.e. million parsecs). In other words, a galaxy 100 million parsecs (326 million light years) distant will be receding at about $100 \times 100 = 10,000$ km. per sec (i.e., increasing its distance by 16,000 miles every second). A galaxy ten times further away will be moving ten times faster, and so on. It seems as if the whole universe is expanding, and that if galaxies exist beyond about 10,000 million light years' range, we cannot see them because they will be moving so fast that light from them will never reach us. This expansion does *not* mean that we are at the centre of the universe; no matter which galaxy we were on we should see the same symmetric picture of receding galaxies.

With visible light, galaxies can be seen out to about 5,000 million light years, but certain galaxies—strong sources of radio waves—can be detected even further away. The objects which we think are most distant of all are *quasars*. Discovered in 1963, they are very compact, looking rather like stars, but have huge red-shifts. If these are interpreted as indicating distance, then some quasars are certainly the most distant objects known and must be radiating up to a hundred times more energy than our galaxy. Yet some of them change in brightness in only a few weeks, which suggests they are quite small. If this is so, then no adequate explanation of their vast energy is known. Some astronomers believe they may not be as distant as we think and controversy rages about them.

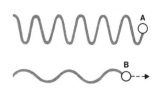

Doppler Effect: the change in wavelength when the source of light is (A) stationary and (B) receding from us

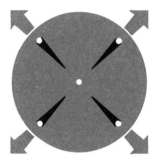

The symmetric recession of the galaxies

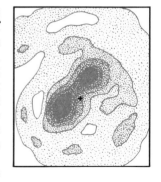

Radio map of a quasar

EVOLUTION OF
THE UNIVERSE

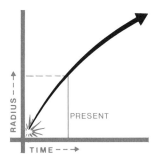

Radius of the universe according to the 'Big Bang' theory

The radius of the oscillating Universe varies periodically

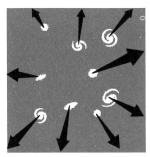

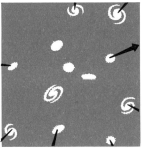

The Steady-State Universe: as old galaxies move apart (above), new galaxies form to take their place (below)

The expansion suggests that the universe is changing and evolving. If the galaxies are expanding away from each other they must have been closer together in the past, and if we go far enough back in time they must have been clumped tightly together. Hubble's Constant suggests that this time was about 10,000 million years ago. *Cosmology* concerns itself with theories of the nature of the universe, and one of these, the *Big-Bang* theory, suggests that about 10,000 million years ago all the material in the universe was lumped together at incredible density, then exploded, forming a rapidly expanding fireball. As expansion continued, temperature fell and galaxies were able to form from the hydrogen gas which was the principal constituent of the universe. The recession of the galaxies which we see now is the velocity left over from the big-bang, and the universe may continue to expand forever.

A variation on this theory, the *Oscillating Universe*, suggests that the expansion may be slowing down and will eventually cease, possibly in about 30,000 million years' time. The universe will then begin to contract until it collapses back to its primitive state. This may be followed by another "Big-Bang", and the universe may continue to pulsate in this fashion.

The value of Hubble's Constant adopted prior to a major revision in 1952 appeared to indicate that the Big-Bang theory could not be true (it seemed as if the "age" of the universe was less than that of some stars!) and an alternative was proposed— the *Steady-State* universe. Basically, this suggested that as the galaxies expanded away from each other, new material was created from which new galaxies formed. The overall view of the universe has always been the same and would remain the same in the future.

Astronomers are trying to sort out which, if any, of these theories is correct. If there were an alternative explanation of the galaxies' red-shifts, then the whole idea of the expansion of the universe might be abandoned. But no other satisfactory explanation exists and we must accept the expansion at its face value for the time being. If the Big Bang is correct, galaxies must have been packed closer together in the past and when we observe distant systems we are seeing them as they use to be thousands of millions of years ago, when their light set out towards us. Consequently, counts of the numbers of distant galaxies should tell us whether or not the universe is evolving. The evidence currently suggests that it is, but is based upon measurements of radio galaxies and quasars. Since we don't know the true nature of these objects the results may be misleading. Recent microwave measurements have detected what may be radiation left over from the Big-Bang, but although the Steady State idea is out of favour the debate is certainly not settled yet.

CHAPTER SEVEN
EXPLORING THE UNIVERSE

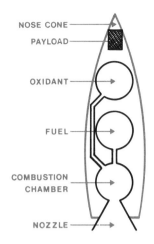

Structure of a chemical rocket

The rocket is the only kind of vehicle which can operate in the vacuum of space, other propulsion systems—such as the airscrew—needing air in which to function. The principle of the rocket is beautifully simple and hinges on Newton's Third Law: "To every action there is an equal and opposite reaction". The rocket burns fuel and produces large quantities of hot gases which would cause it to explode if they did not escape. These gases are allowed to escape through a nozzle at the base of the rocket, and as they rush out in one direction so the rocket shoots off in the opposite direction, just as an inflated balloon will do when its nozzle is released. Present-day rockets use chemical fuels together with an oxidant which causes the fuel to burn at a controlled rate in the *combustion chamber*.

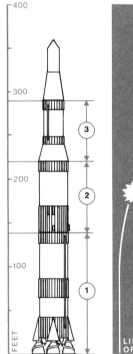

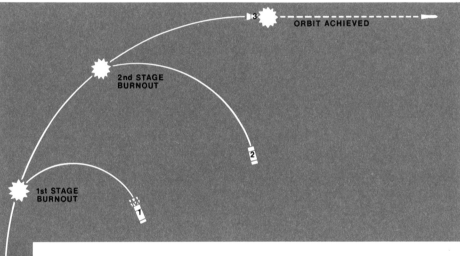

A three-stage rocket: as the fuel in each stage is consumed, it drops away. The third stage places the payload in the planned orbit

The final speed attained by a rocket depends upon the ratio of its initial mass (when full of fuel) to its final mass (when all its fuel is consumed). The bigger this ratio the faster it will go, but since a higher proportion of the rocket then consists of fuel the useful payload must become a lower proportion. To overcome this problem, rockets usually consist of several *stages* on top of each other. When each successive stage runs out of fuel it is dropped off, and the next stage fires. In this way the dead weight of the stage that has used all its fuel does not have to be carried any further, so the final stage of the vehicle can more easily attain high velocities.

1 Too slow: satellite
  falls down

2 Faster, but still too slow

3 Too fast, flies off into
  space

4 Just right (5 miles per
  second), enters circular
  orbit

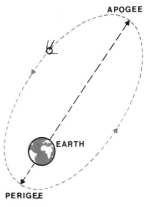

Elliptical orbit

Artificial earth satellites are now commonplace, but the way they stay in orbit is nevertheless of some interest. If we were to stand on top of a high tower and release a stone, it would drop vertically to the ground due to the gravitational pull of the earth which attracts objects towards the earth's centre. If we threw the stone forward—parallel to the ground—it would fall some distance away, and if we threw it harder it would travel further before reaching the ground. If we could throw it at a speed of about 5 miles (8 km.) per second, in a direction exactly parallel to the horizon, then the stone would continue to travel round the earth, never getting nearer the surface. This may seem strange, but although the earth's gravity is all the time attracting the stone towards the centre of the earth, because the stone is moving sufficiently fast all that happens is that its path is bent into a circle with the earth at the centre. This speed is called *circular velocity* and is the minimum speed necessary for a satellite to remain in orbit close to the earth. The value of circular velocity decreases further from the earth, so that the moon, for instance, needs only to move at about 0·56 miles (0·9 km.) per second in its orbit. If a satellite is given a velocity somewhat greater than circular velocity it will enter an elliptical orbit round the earth, the point of closest approach being *perigee* and the most distant point *apogee*. The first artificial satellite was Sputnik I, launched in October 1957.

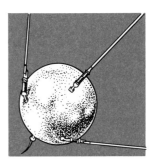

Sputnik I

USES OF
SATELLITES

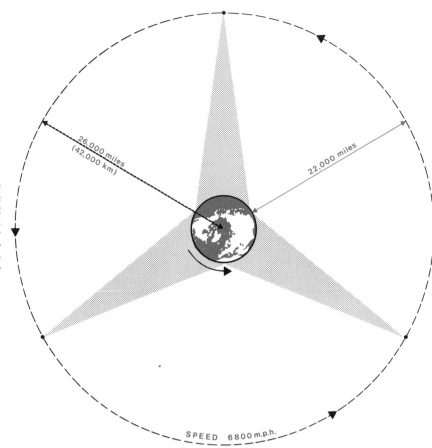

Geosynchronous communications satellites maintain constant positions over the earth and relative to each other: areas covered by their transmissions are indicated by the shading

26,000 miles
(42,000 km)

22,000 miles

SPEED 6800 m.p.h.

Telstar

Orbiting astronomical observatory

Although radio signals can be sent around the world by "bouncing" from a layer in the atmosphere called the *ionosphere*, television signals cannot. Artificial satellites equipped with receiving and transmitting devices and stationed at high altitudes can, however, overcome the problem. Many such *communications satellites* have been established in *geosynchronous* orbits; that is to say they move in circular orbits of radius 26,000 miles (42,000 km.) with orbital periods of 24 hours, and as the earth rotates in the same time they remain permanently above the same positions on its surface. With only three satellites arranged in this way, virtually global television coverage could be achieved. One of the early communications satellites was *Telstar* which relayed the 1964 Olympic games from Tokyo.

Satellites have many uses, e.g., surveying the earth's resources or mapping weather systems, and they are also used astronomically. *Explorer I*, launched by the U.S.A. in 1958, discovered that the earth is surrounded by a belt of charged particles, the first of the Van Allen radiation belts. *Orbiting Astronomical Observatories* (e.g. OAO 2, which carried eleven telescopes) have produced valuable results, and currently great advances are being made in satellite observations of ultra-violet and X-ray sources.

In January 1959 the Russian space probe *Luna 1* passed within 4,000 miles (6,400 km.) of the moon and eight months later *Luna 2* crashed on to the lunar surface. In October 1959 *Luna 3* passed round the far side of the moon, sending back pictures of the hitherto unseen surface. Since then, the moon has been investigated by a succession of different types of spacecraft, such as the American *Ranger* series which send back television pictures prior to crashing into the moon. Later probes soft-landed on the surface, the first being *Luna 9* in 1966, while others, such as the American *Orbiters*, mapped the surface in minute detail from altitudes of only tens of miles.

In recent years, the American approach has been to use manned vehicles of the *Apollo* Series to investigate the surface and bring back to earth rock samples selected by the astronauts. The first manned vehicle to orbit the moon was Apollo 8 in December 1968, while in July 1969 Apollo 11 landed in the Sea of Tranquillity, and Neil Armstrong became the first man to set foot on another world. Experiments were set up on the moon to measure "moonquakes" to see what internal activity was going on in its interior. Subsequent Apollos have followed a similar pattern, sampling different regions of the moon. Apollo 13 suffered a severe explosion in space, but fortunately was able to return safely to earth, while Apollo 15 carried a self-propelled vehicle, the Lunar Rover, for the astronauts' use.

The Russians have continued with unmanned spacecraft. In 1970 *Luna 16* landed, drilled out a small sample of material and returned to earth, as did *Luna 20* in 1972. Luna 17 landed in 1970, together with the automatic roving vehicle Lunokhod 1, which could be controlled by an operator on the earth. The success of this device opened a new chapter in exploration.

To reach the moon by the Apollo method, the spacecraft is first placed in close orbit round the earth (a *parking orbit*). The motors are then fired to move the craft into a long ellipse, its maximum distance from the earth corresponding to that of the moon. As the vehicle moves away, it's slowed down by the gravitational pull of the earth until it reaches a point (the *neutral point*) some 210,000 miles (336,000 km.) away, where the moon's gravity balances that of the earth. Thereafter, the space-craft is accelerated by the moon up to a speed of about 5,750 miles (9,200 km.) per hour and the motors must again be fired to bring the craft into orbit around the moon. Thereafter part of the vehicle, the *lunar module*, takes two astronauts to the surface, while the third remains in orbit in the *command module*. Later, the two astronauts rejoin the command module, shed the lunar module and accelerate back towards the earth. They re-enter our atmosphere at just under 25,000 miles (40,000 km.) per hour, using friction to slow down prior to parachuting into the sea.

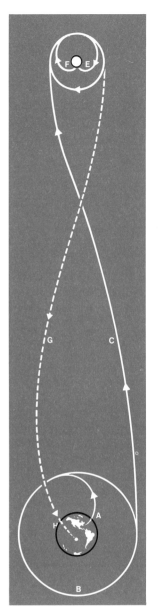

APOLLO LUNAR
FLIGHT TRAJECTORY
A launch, B parking orbit,
C translunar flight,
D lunar orbit,
E lunar landing,
F lift-off and rendezvous
with command module,
G transearth flight,
H splashdown

TO MARS
AND VENUS

In order to reach the planets a spacecraft must be given a speed greater than the earth's *escape velocity*. Recalling the analogy (page 57) of throwing a stone, if we throw an object away from the earth faster than 5 miles (8 km.) per second, we have seen that it will enter elliptical orbit around the earth; but if we throw it faster than 7 miles (11 km.) per second the stone will move away from the earth (along a curve called a hyperbola) and never return, continuing to move away forever without any further use of power. Escape velocity is this critical speed of 7 miles per second.

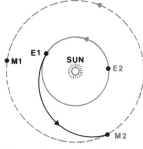

Transfer orbit to Mars

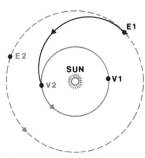

Transfer orbit to Venus

It might seem that the simplest way to reach Mars and Venus, or any other planet, is to fire a rocket straight at them. In principle this could be done if we had sufficiently powerful rockets, or could continue to keep the motors firing all the way. Unfortunately, present-day rockets are not nearly powerful enough to do this, and in any case these methods would be terribly wasteful of fuel. To reach Mars we make use of the earth's velocity in its orbit around the sun, about 18·6 miles (29·8 km.) per second, or about 66,000 miles per hour. The rocket is fired in the same direction in which the earth is moving (see illustration, E1), and after it has escaped from earth it's moving relative to the sun at a speed greater than that of our planet. We can think of the space probe, then, as a tiny planet which will move round the sun, in accordance with Kepler's Laws, in an elliptical orbit that takes it out to, or beyond, the orbit of Mars. The launch must be timed so that the space probe and Mars reach the point M2 at the same time. To reach Venus, the rocket must be fired in the opposite direction so that its velocity subtracts from that of the earth, and it moves on an ellipse that takes it nearer the sun than the earth to intersect the orbit of Venus at V2. Such orbits are called *transfer orbits* and require long flight times (e.g., about 100 days to Venus and 200 days to Mars) from the earth, but are the only method we can use with existing rockets.

Mariner series spacecraft

The first spacecraft to investigate another planet was Mariner 2 which passed within 22,000 miles (35,000 km.) of Venus in 1962, sending back information by radio. The Russians had several unsuccessful attempts, but their probe, Venera 4, landed on the planet in 1967, just one day before the American Mariner 5 flew by at a range of 2,500 miles (4,000 km.). Further Russian spacecraft have subsequently landed there. Mars has also been examined by Russian and American probes. Mariner 4 was the first to send back pictures and other data as it flew by in 1964. Mariners 6 and 7 repeated the exercise in 1969, and Mariner 9 entered Mars orbit in November 1971, to be followed soon after by the Russian vehicles Mars 2 and 3, at least one of which dropped a capsule onto the surface.

60

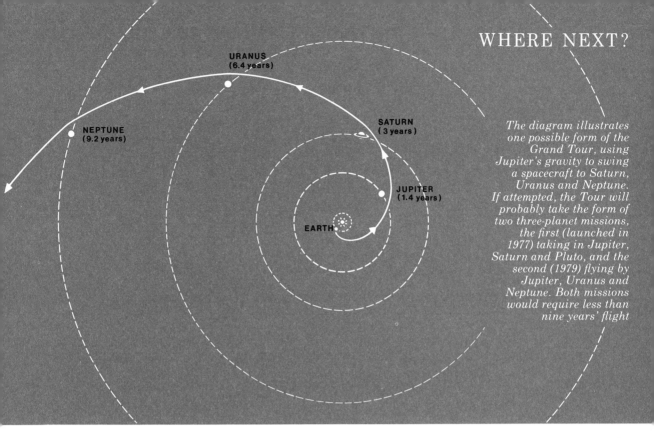

URANUS
(6.4 years)

NEPTUNE
(9.2 years)

SATURN
(3 years)

JUPITER
(1.4 years)

EARTH

The diagram illustrates one possible form of the Grand Tour, using Jupiter's gravity to swing a spacecraft to Saturn, Uranus and Neptune. If attempted, the Tour will probably take the form of two three-planet missions, the first (launched in 1977) taking in Jupiter, Saturn and Pluto, and the second (1979) flying by Jupiter, Uranus and Neptune. Both missions would require less than nine years' flight

Future developments in space exploration will depend upon the finance available. Mars is due to be investigated by further orbiting and landing vehicles, while in March 1972 the first of two small Jupiter probes (*Pioneer 10*) was launched on its two-year mission. Jupiter is a major target for further investigation, an orbiting vehicle being planned; and because it has a strong gravitational field various missions have been devised to make use of this.

Jupiter's gravitational pull can be utilised to accelerate a space probe and change its course, so that it can increase its velocity without using any fuel. Between 1977 and 1981 the four outer planets, Jupiter, Saturn, Uranus and Neptune, will be lined up in such a way that one spacecraft could fly by all of them provided it made use of Jupiter's gravity to assist it. This project, and its variants, has come to be known as "The Grand Tour", but may not be supported in the U.S.A. for financial reasons. Venus will be used in 1973–4 to accelerate a probe to Mercury and a further possibility is that Jupiter's gravity will be employed to send a spacecraft to the sun. To fire a probe direct to the sun requires far more energy that existing rockets can supply, but by passing close to Jupiter a probe could be deflected sufficiently to reach its target. To sum up, it seems as if every major body in the solar system will be investigated by unmanned probes before the year 2000.

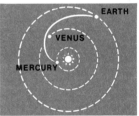

Venus-Mercury probe

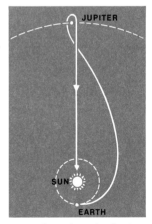

Sun-shot via Jupiter

OTHER PLANETARY SYSTEMS

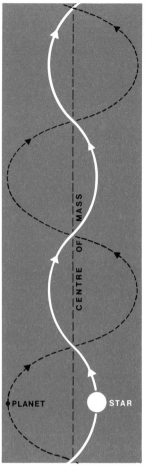

Motion through space of a star with a massive planet

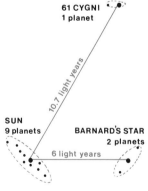

61 CYGNI
1 planet

10.7 light years

SUN
9 planets

BARNARD'S STAR
2 planets

6 light years

Nearby stars thought to have planets

If men ever travel to the stars, they will wish to visit those with planetary systems, since they could not land on the hot surfaces of stars! There is good evidence to suggest that a considerable proportion of stars have planets. We can't see planets of other stars directly—they are far too faint and too distant to be visible even in the largest telescopes—but massive planets can be detected by their gravitational effects on their stars. Just as the earth and moon move round the barycentre (see page 28) so a star and its planet will move around their common centre of mass. A star with such a planet will be seen to wobble slightly as it moves through space, and analysis of this motion tells us something about the planet or planets responsible.

Only planets comparable with Jupiter can be found in this way, but it's reasonable to assume that stars which possess giant planets are likely to have smaller ones too. *Barnard's Star*, about six light years distant, has been analysed by Peter van de Kamp in the U.S.A. and is thought to have two planets comparable to Jupiter. The first star to have its distance measured, 61 Cygni, may have a planet many times larger than Jupiter, and other nearby stars are strongly believed to have planetary systems. As this method is difficult to apply, the fact that several planets have been detected in this way suggests that there may be many planetary systems to be found.

To assess the number of stars in our galaxy likely to have planetary systems we need to know how such systems are formed. The question of the origin of the solar system is not yet settled and there are several possible theories, some suggesting its formation was due to a rare cosmic accident and others that planets arose as a natural consequence of the sun's evolution. The former approach gained popularity through Sir James Jeans who suggested that the material from which the planets formed was dragged from the sun by a passing star. There are well-founded objections to this idea, although it may be possible that the sun could have dragged matter from a passing star rather than vice-versa. If such theories are true, since close encounters between stars are rare, planets must be very rare phenomena. The majority of astronomers feel that the planets formed by condensing from a cloud of material which once surrounded the sun. The cloud was either the debris left over from the formation of the sun itself (or may have been picked up later), or perhaps was ejected from the sun (as was first suggested by Pierre Laplace in 1796) as a result of the rapid rotation it was likely to have had as it contracted from a gas cloud. If these ideas are correct, then stars similar to the sun are very likely to have planetary systems, and there may thus be millions of such systems in our galaxy.

What are the chances of finding life elsewhere in the universe? We don't fully understand the nature of life on our own planet and can say little about totally different forms of life which might be able to exist under conditions possibly totally hostile to us. If we consider only life as we know it, then we require temperatures within a certain range, an atmosphere containing oxygen (and thick enough to shield us from harmful radiations), together with water.

ESSENTIAL ELEMENTS
FOR OUR TYPE OF LIFE:

*Carbon, Oxygen,
Nitrogen, Hydrogen*

The Ecosphere

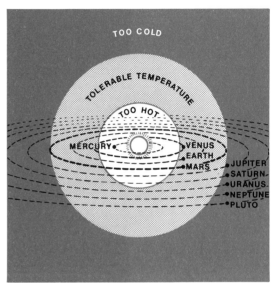

As regards temperature, there is a region in the solar system within which the earth could be moved and at least some of our life forms would continue to exist. This is known as the *ecosphere* and stretches approximately from the orbit of Venus to that of Mars. Though both these planets have unsuitable atmospheres we can't rule out the possibility that some life form may have gained a hold on these planets and evolved in a different way from our own. Otherwise, the chances of finding life on Mars and Venus are extremely slight. Of the other planets, Mercury is too hot, and in any case is airless, while the outer planets are too cold, with quite hostile atmospheres. Even so, suggestions have been made that perhaps a form of life which preferred liquid ammonia to water might exist in Jupiter's atmosphere!

As we have seen, it's likely that thousands of millions of stars have planets, and many of these must lie within the ecospheres of their stars. Of these, some must have suitable conditions, and life will almost certainly have evolved. If we accept this, the possibility of intelligent civilisations elsewhere in the galaxy follows, and there may be others considerably more advanced than ourselves. If such civilisations exist, how could we communicate with them? At present, only radio signals can cross the immense distances to the stars, and since we know that hydrogen gas emits radio waves of 21 cm. wavelength and can assume that other intelligent civilisations would know this too, this would be the logical wavelength at which to transmit to give the best chance of detection. Attempts have been made to pick up intelligent signals from space, but so far without success. Travel to the stars will require some totally different form of propulsion as our present rockets would take millions of years. No doubt it will come in the future.

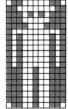

COMMUNICATION WITH
OTHER CIVILISATIONS

*A series of long and short
radio pulses (above)
could be decoded by
elementary mathematical
principles into a
meaningful picture
(below)*

63

INDEX